Frederic Morell Holmes

Firemen and their Exploits

With some Account of the Rise and Development of Fire-Brigades, of Various

Appliances for Saving Life at Fires and Extinguishing the Flames

Frederic Morell Holmes

Firemen and their Exploits
With some Account of the Rise and Development of Fire-Brigades, of Various Appliances for Saving Life at Fires and Extinguishing the Flames

ISBN/EAN: 9783337250089

Printed in Europe, USA, Canada, Australia, Japan

Cover: Foto ©berggeist007 / pixelio.de

More available books at **www.hansebooks.com**

FIREMEN
AND THEIR EXPLOITS:

WITH SOME ACCOUNT

OF THE RISE AND DEVELOPMENT OF FIRE-BRIGADES,
OF VARIOUS APPLIANCES FOR SAVING LIFE
AT FIRES AND EXTINGUISHING
THE FLAMES.

BY

F. M. HOLMES,

AUTHOR OF "ENGINEERS AND THEIR TRIUMPHS," "MINERS AND
THEIR WORKS UNDERGROUND," ETC.

LONDON:

S. W. PARTRIDGE & CO.,

8 & 9, PATERNOSTER ROW.

1899.

PREFACE.

THE present volume, though complete in itself, forms one of a series seeking to describe in a popular and non-technical manner the Triumphs of Engineers. The same style has, therefore, been followed which was adopted in the preceding volumes. The profession of Engineering has exercised great influence on the work of Fire Extinguishment, as on some other things ; and the subject is, therefore, not inappropriate to the series of books of which the volume forms part.

The story of the Fire-Engine begins in Egypt about a hundred and fifty years before Christ. Hero of Alexandria describes a contrivance called the " siphon used in conflagrations," and some persons are of opinion that he was not unacquainted with the use of the air-chest. But it was not until nearly two thousand years later—that is, about the close of the seventeenth century—that the air-chamber and the hose seem to have been brought into anything like general use,—if, indeed, the use can be called general even then.

Much of the story is involved in obscurity, or it may be there was little story to tell; but by the year 1726, Newsham had constructed satisfactory fire-engines in London; and Braithwaite the engineer—who with Ericsson constructed the "Novelty" to compete with Stephenson's "Rocket" at the locomotive contest at Rainhill in 1829—built a steam fire-engine about 1830, though it was not until thirty years, or more, later that the use of the machine became general.

As to Fire-Brigades, the Insurance Companies, which began to appear after the Great Fire of 1666, were wont to employ separate staffs of men to extinguish fires; but by the year 1833, the more important had united, and the London Fire-Brigade had been formed under the control of Mr. James Braidwood. Many provincial towns followed the metropolitan model in forming their brigades.

Together with the development of the Fire-Engine and of efficient brigades has been the introduction of various other appliances, such as Fire-Escapes, Chemical Extinctors, Water-Towers, and the great improvement in the water supply. Nothing is more striking in the history of conflagrations than the comparison between the dry state of the New River pipes at the Great Fire of 1666 and the copious flood of five million gallons poured into the city in a few hours by the same company to quench the great Cripplegate fire of November, 1897.

But, indeed, the whole realm of Fire Extinguishment is a world of constant improvement and strain after perfection. To describe something of these efforts, and trace out the main features of their story, is the object of the present volume.

CONTENTS.

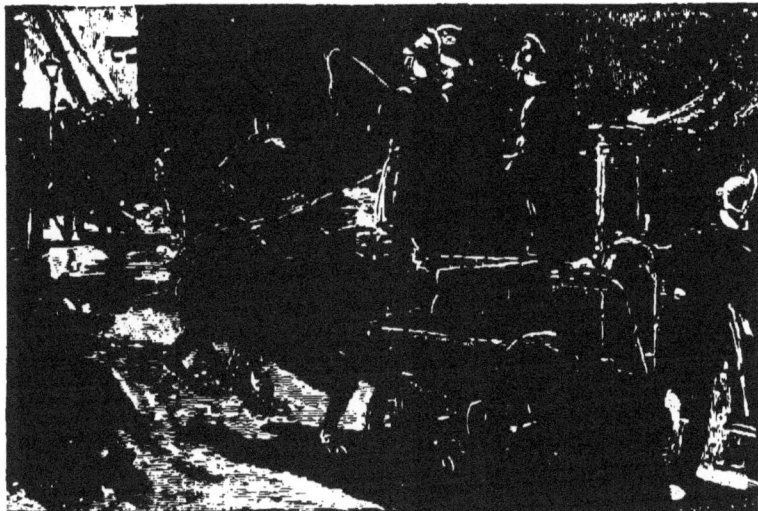

OFF TO THE FIRE.

FIREMEN AND THEIR EXPLOITS.

CHAPTER I.

THE HORSED FIRE-ESCAPE APPEARS. AN EXCITING SCENE.

" SHALL we have a quiet night, Jack ? "
" Can't say," replied Jack philosophically ; " I take it as it comes."

Clang !

Even as he spoke, the electric fire-alarm rang through the silent station. The men sprang toward the stables, glancing at the bell-tablet as they ran.

The tablet revealed the name of the street whence the alarm had been sounded ; and at the clang the horses tossed their heads and pawed the ground, mad

to be off.   They knew the sound of the alarm as well
as the men themselves.

"Will it be a life-saving job, d'ye think, mate?"

"May be," was Jack's sententious reply; "you
never know."

The horses were standing ready harnessed, and were
unloosed at once.   They were led to the engine, the
traces hooked on, the crew, as the staff of firemen
is called, took their places, and the doors in front of
them were opened smartly by rope and pulley.

"Ready?"

"Aye, aye, sir!"

"Right away!"

In less than two minutes from the ringing of the
alarm, the engine was rushing out of the station,
and tearing along London streets with exciting clatter,
the firemen shouting their warning cry, and sparks
flying from the funnel.   Soon the engine fire was
roaring below, and the steam was hissing for its
work.

How had the firemen obtained a blazing fire and
hot steam so soon?   When the engine was waiting
in the station, a lighted gas-jet, kept near the boiler,
maintained the water at a high temperature; and
while the horses were being hooked on, a large fusee,
called a "steam-match," had been promptly ignited,
and dropped flaming down the funnel.   The match fell
through the water-tube boiler to the fuel in the fire-
box below; the draught caused by the rush of the
engine through the air helped the fire; and the water
being already so hot, steam pressure soon arose.

"The new escape's close behind!" cried one of the
men, as the engine hurried along.

Something, unusual then, to London streets was

rapidly following the steamer. In the gloom, it looked like a dim spectral ladder projecting over the horses in front, and several men could be seen sitting on the carriage conveying it.

"She's a-comin' on pretty fast," exclaimed one of the men; "she travels as smart as an engine."

Indeed, the new escape was now so near, that it could be seen more clearly. It was securely mounted on a low car, and its large wheels hung over the end at the back, not far above the ground. Designed by Commander Wells, chief officer of the London Fire-Brigade, it was brought into use in the brigade in July, 1897.

But now it was nearing the fire, and cheers and cries rang loudly from the excited crowd gathered at the spot.

"Make way for the escape! Hurrah! Hurrah!"

No wonder the crowd were excited. On the second-floor window of a large building appeared three white, eager faces, framed by the dark sashes, and crying eagerly for help.

Cheer after cheer rent the air, as the escape drew up opposite, and was slipped from its car; then, resting on its own wheels, it was pitched near the burning building, and its ladders run up to the window. The policemen could scarce keep back the thronging crowd.

Away go the firemen up the rungs of the ladder, and amid continued cheers, and cries, and great excitement approach the sufferers in their peril.

"They've got one!" shouts an excited voice.

"Aye, and there's another!" cries a second spectator.

"They're all three saved!" vociferates a third; and loud cheers greet the firemen's triumph.

It was a smart piece of work ; and with the rescued persons thrown over their shoulders in the efficient manner they are taught at drill, the firemen carefully descend the ladder one after the other, and amid shouts and plaudits arrive safely on the ground.

The flames dart out of the building more fiercely than ever, as if in anger at losing their prey ; the glare and heat grow more intense ; the smoke rolls off in dense volumes ; the fire is raging furiously.

Engine after engine rushes fast to the spot, the loud, alarming cries of " Fire-ire ! Fire-ire ! " echoing shrilly along the lamp-lighted thoroughfares ; fireman after fireman leaps from the arriving engines, and with their bright brass helmets flashing in the glare are quickly stationed round the huge conflagration.

The " brigade call " has been telephoned all round London, and from east and west, and north and south, engines and firemen have hurried to the spot. Steamers with sparks flying, steam hissing, and whistles shrieking ; manuals with the clatter of their handles ; hose-carts with their lengths of flexible pipes ; and tall ladders of fire-escapes, useful, even when no life is to be saved, as high points of vantage whence firemen can direct streams of water straight into the raging fire,—all—all are here.   One after another they arrive, until the word is passed that more than twenty engines and a hundred and twenty firemen are concentrated on the spot.

Hydrants also are at work.   They are appliances, permanently fixed under the pathway, from which firemen can obtain a powerful pressure of water, ranging from thirty-five to seventy pounds per square inch.   From the steamers and the hydrants the quantity of water poured on the huge fire is now immense, and the steam and smoke roll off in immense volumes.

Crash !

"There goes the glass!" cries a fireman; and a few moments later it is rumoured that one of the brigade has been badly cut in the hands. The skylight had broken and fallen upon him, showing that it is not only from heat and smoke that the men are likely to suffer, but also from falling parts of the burning building.

The huge fire is fought at every possible point. It is prevented from spreading to surrounding buildings by deluging them with water, and strenuous efforts are made to quench it at its source. Steadily in the growing light of day the firemen work on; but the morning had far advanced before the great conflagration was fully extinguished and the London Salvage Corps were left in possession of the ruined premises.

"Well, you've had your first big fire, Newall; how d'ye like it?"

"Oh, it's all right, mate; it's pretty hard work, but I don't mind it."

"'Tain't all over yet," said Jack cheerfully; "there's this 'ere hose to be scrubbed and cleaned, and hung up in the well to dry. I reckon it will be four or five o'clock before we can turn in."

Jack was right. The wet hose had to be suitably treated to keep it in good condition, and the engines carefully prepared for the next alarm that might arise; and when the men turned in to rest, they slept sound enough.

This story not only illustrates the work of the London Fire-Brigade, but also points to a notable fact

in its history. That fact is the introduction of the
horsed fire-escape. The first rescue in London by
this valuable appliance took place on October 17th,
1898. There were, in fact, two disastrous fires raging
at nearly the same time on that day, and the new
appliance was used at one of these.

Early in the morning, a disastrous fire broke out in
Manresa Road, Chelsea. The conflagration originated
in the centre of a large timber-yard, and spread so
rapidly that a very serious fire was soon in progress.
Engines and firemen hurried up from various quarters,
until sixteen steamers, three manuals, and more than
a hundred men were on the spot. The fire was completely surrounded, and the enormous quantity of water
poured upon the blazing wood soon took effect.

But before all the engines had left, news came that
a still more serious fire had broken out in Oxford
Street. The extensive premises of Messrs. E. Tautz
& Co., wholesale tailors, were discovered to be in
flames, and the alarm was brought to the fire-stations
from various sources.

The Orchard Street fire-alarm rang into Manchester
Square station, and resulted in the horsed escape
being turned out; then another fire-alarm rang
into Great Marlborough Street fire-station, and the
horsed escape had hurried from this point also. The
appliance was new, and for some time the men of
the brigade had cherished a laudable ambition to be
the first to use the escape in what they call a life-
saving job. And it was only by an untoward chance,
or simple fortune of war, that the men of the
Manchester Square station, who were first on the
spot, missed the coveted honour.

When they arrived on the scene, no sign of fire was

visible in Oxford Street itself, and the firemen were pointed to North Row, one of the boundaries of the burning block behind. They made their way thither, searching for inmates, but were driven back by the fierce flames.

Meantime, the three persons sleeping on the premises—the foreman, Mr. Harry Smith, his wife, and their little son, aged six years—had been endeavouring to escape by the staircase, but had been driven back by the fire. Mr. Smith had been awakened by the dense smoke filling the room, and he aroused his wife at once and took the boy in his arms.

Not being able to escape by the staircase, they hurried to the front of the large block of buildings, shutting the doors after them as they went. So it happened that they appeared at the second-floor windows facing Oxford Street just as the horsed escape from Great Marlborough Street fire-station hurried up. A scene of great excitement followed. The firemen ran the ladders from the escape to the building, and brought down all three persons in safety; but Mrs. Smith unfortunately had suffered a burn on the left leg. It is probable that, but for the rapidity with which the horsed escapes arrived on the scene, the family might have suffered much more severely; for the fire was very fierce, and soon appeared in Oxford Street.

The honour, therefore, of the first rescue by the new horsed escape rests with the Great Marlborough Street station, though the efforts of their brave comrades of the Manchester Square station should always be remembered in connection therewith. Commander Wells appreciated this; for he telephoned a special message to Superintendent Smith, saying :

" Please let your men understand that I thoroughly appreciate and approve their action on arrival at the fire this morning, although the honour of rescue falls by the fortune of war to the second horse-escape."

The fire proved very disastrous, and a large force was speedily concentrated. It was eventually subdued; but it was about two o'clock in the afternoon before the brigade were able to leave, a large warehouse belonging to Messrs. Peel & Co., boot-makers, being also involved, and other buildings more or less damaged.

The horsed fire-escape, which was found so useful on this occasion, is but one among several appliances for saving life and fighting the fire. These appliances are worked by highly-trained brigades of firemen, whose efficient organization, well-considered methods, and ingenious apparatus form one of the remarkable features of the time.

They did not reach their present position in a day. Indeed, a stirring story of human effort and of high-spirited enterprise lies behind the well-equipped brigades of the time. Step by step men have won great victories over difficulty and danger; step by step they have profited by terrible disasters, which have spurred them on to fresh efforts.

What, then, is this story of the fight against fire? How have the fire-services of the day reached their present great position?

CHAPTER II.

THE BEGINNING OF THE STORY. HERO'S "SIPHON."
HOW THE ANCIENTS STROVE TO EXTINGUISH FIRES.

No one knows who invented the modern fire-engine.

The earliest machine, so far as is generally known, was described by Hero of Alexandria about a hundred and fifty years before Christ. He called it "the siphon used in conflagrations"; and it seems to have been originated by Ctesibius, a Greek mechanician living in Egypt, whose pupil Hero became.

It is very interesting to notice how this contrivance worked. It was fitted with two cylinders, each having a piston connected by a beam. This beam raised and lowered each piston alternately, and with the help of valves—which only opened the way of the jet—propelled water to the fire, but not continuously. The method must have proved very inefficient, especially when compared with the constant stream thrown by the modern fire-engine. Indeed, it is this power to project a steady and continuous stream which chiefly differentiates the modern fire-engine from such machines as Hero's siphon.

How far this siphon or any similar contrivance was used in ancient times we cannot say; but no doubt buckets in some form or other were the first appliances used for extinguishing conflagrations. Whenever mankind saw anything valuable burning, the first impulse would be to stamp it out, or quench the flame by throwing water on it; and the water would be conveyed by the readiest receptacle to hand; then when men had discovered the use of the pump, or

2

the squirt, they would naturally endeavour to turn
these appliances to account.

In some places the use of water-buckets was
organized. Juvenal alludes to the instructions of
the opulent Licinus, who bade his "servants watch
by night, the water-buckets being set ready"; the
wealthy man fearing "for his amber, and his statues,
and his Phrygian column, and his ivory and broad
tortoise-shell."

Then Pliny and Juvenal use a term—*hama*—which
signifies an appliance for extinguishing fires; but the
true rendering seems to be in dispute, some trans-
lators being content to describe it simply as a water-
vessel. Pliny the Younger refers to *siphones*, or
pipes, being employed to extinguish fires; but we
do not know how they were used, or whether they
resembled Hero's siphon.

In fact, the earliest references to fire-engines by
Roman writers are regarded by some as being merely
allusions to aqueduct-pipes for bringing water to
houses, rather than to a special appliance. And from
Seneca's remark, "that owing to the height of the
houses in Rome it was impossible to save them when
they took fire," we may gather that any appliances
that may have been in use were very inefficient.

A curious primitive contrivance is described by
Apollodorus, who was architect to Trajan. It consisted of leathern bags or bottles, having pipes
attached; and when the bottles were squeezed, the
water gushed through the pipes to extinguish the
flames. Augustus was so enterprising as to organize
seven bands of firemen, each of which protected two
districts of Rome. Each band was in charge of a
tribunus, or captain, and the whole force was under a

HOW THE ANCIENTS STROVE TO EXTINGUISH FIRES. 19

præfectum vigilum, or prefect of the watch ; though what apparatus they employed—whether buckets or pipe-bags, syringes or Hero's siphon—we do not know.

But these appliances, or some of them, were no doubt in use at the Great Fire of Rome in A.D. 66. In July of that year—the tenth of the reign of the infamous Emperor Nero—two-thirds of the city was destroyed. The fire broke out at a number of wooden shops built against the side of the great Circus, and near to the low-lying ground between the Palatine and the Cælian Hills. The east wind blew the flames onward to the corner of the Palatine Hill, and there the fire blazed in two directions. It gained such enormous power, that stonework split and fell before it like glass, and building after building succumbed, until at one point it was only stopped by the river, and at another by frowning cliffs.

For six awful days and seven nights the fire raged, and then, when it was supposed to have been extinguished, it burst forth again for three more days. The sight must have been appalling. We can picture the huge sheets and tongues of flame sweeping ever onward, the fearful heat, and the immense volumes of smoke which mounted upward and obscured the sky.

The panic-stricken people fled to the imperial gardens, but whispered that Nero himself had originated the fire. To divert suspicion, he spread reports that the Christians were the culprits ; and they were treated with atrocious cruelty, some being wrapped in fabric covered with pitch and burnt in the Emperor's grounds. The guilt of Nero remains a moot point ; but he seems to have acted with some amount of liberality to the sufferers, though his acts

of humanity did not free his name from the foul
suspicion.

The conflagration itself stands out as one of the
most terrible in history.   Before its furious rage the
capable Romans seem to have been reduced to impo-
tence.   Their organization, if they had any, seems to
have been powerless ; and their appliances, if they used
any, seem to have been worthless.

We are entitled to draw the deduction that they
had no machine capable of throwing a steady, continuous stream from a comparatively safe distance.
No band of men, however strong and determined, could
have stood their ground sufficiently near the fierce
fire to throw water from buckets, pipe-bags, or even
portable pumps.   For small fires they might prove of
service, if employed early ; but for large conflagrations they would be worthless.   And if Rome, the
Mistress of the World, was so ill-provided, what must
have been the condition of other places ?

We may infer, therefore, that the means of fire
extinction in the ancient world were miserably
inadequate.

Had mediæval Europe anything better to show ?

CHAPTER III.

IN MEDIÆVAL DAYS.   AN EPOCH-MAKING FIRE.

" Prithee, good master, what's o' fire ? "

" A baker's house they say, name of Farryner."

" Faith ! it's in Pudding Lane, nigh Fish Street
Hill," quoth another spectator, coming up.   " They
say the oven was heated overmuch."

" It's an old house, and a poor one," said another
speaker. " 'Twill burn like touchwood this dry
weather."

" Aye, it have been dry this August, sure enow ; and
I reckon the rain won't quench it to-night." And the
speaker looked up to the starlit sky, where never a
cloud could be seen.

" Have they the squirts at work, good-man ? "

" Aye, no doubt. 'Twill be quenched by morning,
neighbour. Faith ! 'tis just an old worm-eaten house
ablaze, and that's the tale of it."

But it was not " the tale of it." A strong east wind
was blowing, and the hungry flames spread quickly to
neighbouring buildings. These houses were old and
partly decayed, and filled with combustible material,
such as oil, pitch, and hemp used in shipwright's work.
In a comparatively short time the ward of Billingsgate
was all ablaze, and the fierce fire, roaring along Thames
Street, attacked St. Magnus Church at Bridgefoot.

Before the night was far spent, fire-bells were
clashing loudly from the steeples, alarming cries of
" Fire ! Fire ! " resounded through the streets, and
numbers of people in the old narrow-laned city of
London were rushing half dressed from their beds.

It was the night of Saturday, September 2nd, 1666,
a night ever memorable in the history of London.
About ten o'clock, any lingerers on London Bridge—
where houses were then built—might have seen a
bright flame shoot upward to the north. They
probably conversed as we have described, and retired
to bed. But the fire spread from the baker's shop, as
we have seen, and the confusion and uproar of that
terrible night grew ever more apace.

Half-dazed persons crowded the streets, encumbered

with household goods, and the narrow thoroughfares soon became choked with the struggling throng. But the flames seized upon the goods, and the panic-stricken people fled for their lives before the fierce attack. The lurid light fell on their white faces, and the terrible crackling and roaring of the flames mingled with their shrieks and shouts as they hurried along. Now the night would be obscured by dense clouds of thick smoke, and anon the fire would flash forth again more luridly than ever.

To add to the alarm, the cry would ring through the streets, or would be passed from mouth to mouth, that the pipes of the New River Company—then recently laid—were found to be dry. With the suspicion of Romanist plots prevailing, the scarcity of water and the origin of the fire were put down to fanatical incendiaries ; or, as an old writer quaintly expressed it, "This doth smell of a popish design."

When the next morning dawned, the terrible conflagration, so far from having been extinguished, was raging furiously ; the little jets and bucketsful of water, if any had been used, proved of no avail ; and the narrow streets became, as it were, great sheets of flame.

But was nothing done to extinguish the fire ? What appliances would the Londoners have had ?

Here, perhaps, in the early hours of the conflagration, you might have seen a group of three men at the corner of a street working a hand-squirt. This instrument was of brass, and measured about 3 feet long. Two men held it by a handle on each side ; and when the nozzle had been dipped into a bucket or a cistern near, and the water had flowed in, they would raise the squirt, while the third man pushed up the piston to discharge the water. The squirt might hold about four quarts of water.

If one man worked the squirt, he would hold it up

A CITY FIRE TWO HUNDRED YEARS AGO.

by the handles, and push the end of the piston, which was generally guarded by a button, against his chest. But, at the best, it is obvious that the hand-squirt was a very inadequate contrivance.

Not far distant you might also have seen a similar

squirt, mounted in a wheeled reservoir or cistern, the
pistons, perhaps, worked by levers ; and, possibly, in yet
another street you might have noticed a pump of some
kind, also working in a cistern ; while here and there
you might have come upon lines of persons passing
buckets from hand to hand, bringing water either
from the wells in the city, or from the river, or
actually throwing water on the fire. Such were the
appliances which we gather were then used for
extinguishing fires.

But such contrivances as were then in the neighbourhood of Fish Street Hill appear to have been
burnt before they could be used, and the people seem
to have been too paralyzed with terror to have
attempted any efforts.

The suggestion was made to pull down houses, so as
to create gaps over which the fire could not pass ; and
this suggestion no doubt indicates one of the methods
of former days. But the method was not at first
successful on this occasion.

Thus, Pepys, in his Diary, tells us, under date of
the Sunday : "At last [I] met my Lord Mayor in
Canning Street, like a man spent, with a handkercher
about his neck. To the King's message [to pull down
houses before the fire] he cried, like a fainting woman,
'Lord ! what can I do ? I am spent : people will not
obey me. I have been pulling down houses ; but the
fire overtakes us faster than we can do it.'" This is
a graphic little picture of the bewilderment of the
people ; and Pepys goes on to say that, as he walked
home, he saw "people all almost distracted, and no
manner of means used to quench the fire."

In a similar manner, another famous eye-witness,
John Evelyn, notes in his Diary that "some stout

THE GREAT FIRE OF LONDON (FROM A CONTEMPORARY PRINT).

seamen proposed, early enough to have saved nearly
the whole city," the destruction of houses to make a
wide gap ; "but this some tenacious and avaricious
men, aldermen, etc., would not permit, because their
houses must have been of the first."

The main idea, therefore, of extinguishing the fire
seems to have lain in the pulling down of houses to
produce a wide gap over which the fire could not
pass. But at first the civic authorities shrank from
such bold measures. On Sunday, then, the flames
were rushing fiercely onward, the ancient city echoing
to their roaring and to the cries and shrieks of the
populace. The houses by London Bridge, in Thames
Street, and the neighbourhood were but heaps of
smouldering ruins. The homeless people sought refuge
in the fields outside the city by Islington and Highgate, and the city train-bands were placed under arms
to watch for incendiaries ; while, as if the horror of
the terrible fire was not enough, numbers of ruffians
were found engaged in the dastardly work of plunder.
The clanging of the fire-bells, the crackling of the
huge fire, the cries and curses of the people, made
such a frightful din as can scarce be imagined ; while
many churches, attended on the previous Sunday by
quiet worshippers, were now blazing in the fire.

That night the scene was appalling, and yet magnificent. An immense sheet of fire rose to the sky,
rendering the heavens for miles like a vast lurid
dome. The conflagration flamed a whole mile in
diameter, hundreds of buildings were burning, and the
high wind bent the huge flames into a myriad curious
shapes, and bore great flakes of fire on-to the roofs
of other houses, kindling fresh flames as they fell.
For ten miles distant the country was illumined as

at noonday, while the smoke rolled, it is said, for fifty miles.

Evelyn describes the scene in his Diary, under date September 3rd : " I had public prayers at home. The fire continuing, after dinner I took coach with my wife and son and went to the Bankside in Southwark, where we beheld the dismal spectacle, the whole city in dreadful flames near the water-side ; all the houses from the Bridge, all Thames Street, and up-wards towards Cheapside, down to the Three Cranes, were now consumed : and so returned exceeding astonished what would become of the rest.

" The fire having continued all this night (if I may call that night which was light as day for ten miles round about, after a dreadful manner) when con-spiring with a fierce eastern wind in a very dry season ; I went on foot to the same place, and saw the whole south part of the city burning from Cheapside to the Thames and all along Cornhill. . . . Here we saw the Thames covered with goods floating, all the barges and boats laden with what some had time and courage to save, as, on the other [side], the carts, etc., carrying out to the fields, which for many miles were strewed with moveables of all sorts, and tents erect-ing to shelter both people and what goods they could get away. Oh the miserable and calamitous spectacle ! such as haply the world had not seen the like since the foundation of it, nor be outdone till the universal conflagration of it ! All the sky was of a fiery aspect, like the top of a burning oven, and the light seen above forty miles round about for many nights. God grant mine eyes may never behold the like, who now saw above ten thousand houses all in one flame ; the noise and cracking and thunder of the impetuous

flames, the shrieking of women and children, the
hurry of people, the fall of towers, houses, and
churches, was like a hideous storm, and the air all
about so hot and inflamed that at the last one was
not able to approach it, so that they were forced
to stand still and let the flames burn on, which they
did for near two miles in length and one in breadth.
The clouds also of smoke were dismal, and reached,
upon computation, near fifty-six miles in length. Thus
I left it this afternoon burning, a resemblance of
Sodom or the last day."

On Monday the Royal Exchange perished in the
sea of flame. By evening Cheapside had fallen,
and beside the water's edge it was blazing in Fleet
Street; while it had also burned backward, even
against the wind, along the eastern part of Thames
Street, toward Tower Hill. The heat was so terrible
that persons could not approach within a furlong,
while the very pathways were glowing with fiery
heat. Some persons chartered barges and boats, and,
filling them with such property as they could save,
sent them down the Thames. Others paid large sums
for carts to convey property far beyond the city walls.
A piteous exodus of sick and sound, aged and young,
crawled or fled to the spacious fields beyond the gates.
The ground was strewn with movables for miles, and
tents were erected to shelter the burned-out multitude.

At length St. Paul's succumbed. It had stood
tall and strong in the space of its churchyard, lifting
its head loftily amid the billows of flame; but at
last the terrible fire, driven toward it by the east
wind, lapped the roof, and seized some scaffold-
poles standing around. The lead on the roof melted
in the fierce heat, and ran down the walls in

streams ; the stones split, and pieces flew off with reports like cannon-shots ; and beams fell crashing like thunder to the ground.

Evelyn notes, under date September 4th : " The burning still rages, and it was now gotten as far as the Inner Temple ; all Fleet Street, the Old Bailey, Ludgate Hill, Warwick Lane, Newgate, Paul's Chain, Watling Street now flaming, and most of it reduced to ashes ; the stones of Paul's flew like granados, the melting lead running down the streets in a stream, and the very pavements glowing with fiery redness, so as no horse nor man was able to tread on them, and the demolition had stopped all the passages, so that no help could be applied. The eastern wind still more impetuously driving the flames forward. Nothing but the almighty power of God was able to stop them, for vain was the help of man."

On the eastern side of St. Paul's, the old Guildhall fell to the fire. On Tuesday night, it was, says a contemporary writer, the Rev. Thomas Vincent, in a little volume published a year afterwards, " a fearfull spectacle, which stood the whole body of it together in view, for several hours together, after the fire had taken it, without flames (I suppose because the timber was such solid oake), in a bright shining coale as if it had been a Pallace of gold, or a great building of burnished brass."

The fire had now become several miles in circumference. It had reached the Temple at the western end of Fleet Street by the river, and was blazing up by Fetter Lane to Holborn ; then backward, its course lay along Snow Hill, Newgate Street—Newgate Prison being consumed—and so past the Guildhall and Coleman Street, on to Bishopsgate Street and

Leadenhall Street. It seemed as though all London would be burnt, and that it would spread westward even to Whitehall and Westminster Abbey.

But now the King (Charles II.) and his brother the Duke of York and their courtiers were fully aroused ; and it must have become clear to even the meanest intelligence that houses must be blown down on an extensive scale, in order to create large gaps over which the fire could not pass. All through Tuesday night, therefore, the sound of explosions mingled with the roaring of the fire.

By the assistance of soldiers, and by the influence of the royal personages, buildings were blown up by gunpowder in the neighbourhood of Temple Bar, which then, of course, spanned the western end of Fleet Street ; at Pye Corner near the entrance to Smithfield, and also at other points of vantage. These bold means, together, no doubt, with the falling of the wind, and also the presence of some strong brick buildings, as by the Temple, checked and stopped the fire. Some began now to bestir themselves, " who hitherto," remarks Evelyn, " had stood as men intoxicated with their hands across." On the Wednesday, therefore, the fire extended no farther west than the Temple, and no farther north than Pye Corner near Smithfield ; but within this area it still burned, and the heat was still so great that no one would venture near it.

During the Wednesday, the King was most energetic. He journeyed round the fire twice, and kept workers at their posts, and assisted in providing food and shelter for the people. Orders were sent, into the country for provisions and tents, and also for boards wherewith to build temporary dwellings. On Thursday

the Great Fire was everywhere extinguished ; but on
Friday the ruins were still smouldering and smoking,
and the ground so hot that a pedestrian could not stand
still for long on one spot.   From St. Paul's Churchyard, where the ground rises to about the greatest
height in the old city, the eye would range over a
terrible picture of widespread destruction, from the
Temple to the Tower and from the Thames to Smithfield.   Two hundred thousand homeless persons were
camping out, or lying beside such household goods as
'they had been able to save, in the fields by Islington
and Highgate.   It has been computed that no fewer
than 13,200 houses, 89 churches, including St. Paul's,
400 streets, and several public buildings, together
with four stone bridges and three of the city gates,
etc., were destroyed, while the fire swept over an
area of 436 acres.

Now, in connection with this great calamity, we
cannot find any appliance at work corresponding to
our modern fire-engine.   The inhabitants of London
seem to have been almost, if not quite, as badly
provided against fire as Rome in the days of Nero.

In fact, the chief protection in early days in England
seems to have been a practice of the old proverb that
prevention is better than cure, care being exercised
to regulate the fires used for domestic purposes : we
see an instance in the arrangement of the curfew-
bell, or *couvre-feu*, a bell to extinguish all fires at
eight at night.   Still, when conflagrations did occur,
we may suppose that buckets and hand-squirts, as
soon as mankind came to construct them, were the
appliances used.

Entries for fire-extinguishing machines of some
sort have been found in the accounts of many German

towns : for instance, in the building accounts of Augsburg for 1518, "instruments of fire " or " water-syringes " are mentioned.

Fires appear to have been very frequent in Germany in the latter part of the fifteenth and in the sixteenth century. And though we do not know much of the contrivances used in Europe in the Middle Ages, it is not until 1657 that we have any reliable record of a machine at all resembling Hero's siphon on the one hand, or the modern fire-engine on the other.

This record is given by Caspar Schott, a Jesuit, and tells of an engine constructed by Hautsch of Nuremberg, a city long famous for mechanical contrivances. The machine was really a large water-cistern drawn on a wheeled car, or sledge ; and the secret of its propulsive power, Schott supposes was a horizontal cylinder containing a piston and producing an action like a pump. The cistern measured 8 feet long by 4 feet high, and 2 feet wide ; its small width being probably designed for entering narrow streets. It was operated by twenty-eight men, and it forced a stream of water an inch thick to a height of about eighty feet. Hautsch desired to keep the methods of its construction secret ; but, apparently, it was not furnished with the important air-chamber, and does not seem to have differed very materially from Hero's siphon. Schott also says he had seen one forty years before at Königshofen.

Notwithstanding, therefore, the danger of great conflagrations, mankind does not seem to have made much progress in the construction of fire-engines from the days of Ctesibius until the time of Charles II., a period of about eighteen hundred years. On the other hand, we must remember that syringes and

water-buckets can be of very great service when
promptly and efficiently used. Even to-day London
firemen find similar appliances of great value for
small conflagrations in rooms.

But we get a vivid little picture of the helplessness
of even the seventeenth-century public before a fire
of any size, in a description left by Wallington of a
fire on Old London Bridge in 1633. Houses were
then built on the bridge, and Wallington says : "All
the conduits near were opened, and the pipes that
carried the water through the streets were cut open,
and the water swept down with brooms with help
enough ; but it was the will of God it should not
prevail. For the three engines which are such
excellent things that nothing that ever was devised
could do so much good, yet none of them did prosper,
for they were all broken, and the tide was very low
that they could get no water, and the pipes that were
cut yielded but littel. Some ladders were broke to
the hurt of many ; for several had their legges broke,
some their arms ; and some their ribes, and many
lost their lives." More than fifty houses, we may
add, were destroyed by this fire.

Of what character were the engines to which he
refers we cannot tell. We do not know whether any
engine like Hautsch's was established in London at
this time, or at the date of the Great Fire ; but if
so, it was not apparently much in vogue. It must
be remembered that the term "engine" was applied
indiscriminately to any sort of mechanical contrivance,
and even to a skilful plan or method (Shakespeare
uses the word to designate an instrument of torture) ;
if, therefore, the word is used for a fire-extinguishing
appliance by any old writer, it does not follow that

the so-called engine would resemble Hautsch's machine or a modern fire-engine.

Judging from some Instructions of the Corporation after the fire, hand-squirts and ladders and buckets were still chiefly relied upon in 1668.   The Instructions

FIRE-EXTINGUISHING APPLIANCES, SQUIRTS,
BUCKETS, ETC., A.D. 1667.

are, moreover, interesting, as showing what action the Corporation took after the Great Fire.

The city was divided into four districts, each of which was to be furnished with eight hundred leathern buckets, fifty ladders varying in sizes from 16 to 42 feet long, also " so many hand-

squirts of brass as will furnish two for every
parish, four-and-twenty pickaxe-sledges, and forty
shod shovels." Further, each of the twelve companies was to provide thirty buckets, one engine,
six pickaxe-sledges, three ladders, and two hand-
squirts of brass. Again, "all the other inferior
companies" were to provide similar appliances; and
aldermen were likewise to provide buckets and hand-
squirts of brass. The pickaxes and shovels were for
use in demolishing houses and walls if necessary, or
dealing with ruins; and though some kind of engine
is mentioned, we know not whether it was a hand-
squirt mounted in a cistern, or some sort of portable
pump.

We may regard these regulations, however, as
fixing for us the hand-squirt and the bucket as the
principal means of fire extinguishment in Britain
up to that date.

But now a great development was at hand, and
a new chapter was to commence in the story.

CHAPTER IV.

THE PEARL-BUTTON MAKER'S CONTRIVANCE. THE
MODERN FIRE-ENGINE.

How to force a continuous stream of water on the
fire !

That was the problem which puzzled an unknown
inventor about the year 1675. He probably saw
that hitherto the appliances for extinguishing conflagrations failed at this point, and we may suppose

that he cudgelled his brains to hit upon the right remedy.

Then one day, no one seems to know when, he thought of inventing, or adapting, the compressed air-chamber to a sort of portable pump, and, behold !— The Modern Fire-Engine was born !

The invention was introduced, probably, after the Great Fire, because authorities describe it as first mentioned in the French *Journal des Savans* in 1675, and Perrault states that an engine with an air-chamber was kept at Paris for the protection of the Royal Library in 1684. If, therefore, Hero knew of the air-chamber, as some assert, it does not appear to have been much used. But probably the great disaster in London stirred invention, and the addition of the air-chamber was the result. It may not, however, have been a distinct invention, for an air-chamber had been found of great value in various hydraulic machines.

What, then, is this invention, and what is its great value to a fire-engine ?

Briefly, it enables a steady and continuous stream of water to be thrown on a fire. It is the vital principle of the modern fire-engine, and renders it distinctly different from all squirts, syringes, and portable pumps preceding it. Instead of an unequal and intermittent supply, sometimes, no doubt, falling far short of the fire, we have now a persistent stream, which can be continuously directed to any point, in reach, with precision and efficiency.

How, then, are these results obtained ? How does the air-chamber work ?

It depends on the elasticity and power of compressed air. The water, when drawn from the source of supply

by two pistons, working alternately, is driven into
a strong chamber filled with air. The air becomes
compressed, and is driven to one part of the chamber ;
but when it is forced back to occupy about one-third
of the whole space, the air is so compressed that,
like the proverbial worm which will turn at last,
it exerts a pressure on the water which had been
driving it back. If the water had no means of
escape, the chamber would soon burst ; but the water
finds its way through the delivery-hose. If the hose
issue from the top of the chamber, it is fitted with
a connecting pipe reaching nearly to the bottom to
prevent any escape of air.

Now, as long as the pumps force the water into the
air-chamber to the necessary level—that is, to about
two-thirds of the space—the pressure is practically
continuous, and thus a constant jet of water is
maintained through the hose. The ordinary pressure
of air is about 14·7 pounds per square inch ; and when
compressed to one-half its usual bulk, its elasticity
or power of pressure is doubled, and of course is
rendered greater if still further compressed.

This power, then, of the compressibility and elasticity
of air is the secret of the fire-engine air-chamber ; but
though introduced about 1675, it was not until 1720
that such engines seem to have become more general.
About that date, Leupold built engines in Germany
with a strongly-soldered copper chest, and one piston
and cylinder, the machine throwing a continuous and
steady jet of water some twenty or thirty feet high.

In the meantime, what was being done in England ?

Here again the story is obscure ; but we imagine
the course of events to have been something like this :

In the dismal days after the Great Fire, people

began to cast about for means to prevent a recurrence
of so widespread and terrible a calamity. Fire-insurance offices were organized, and they undertook the
extinguishment of fires. It is not unreasonable to
suppose that in some form—perhaps by offering
prizes, perhaps by simply calling attention to the need
for improvement, perhaps by disseminating information
such as of the engine mentioned by Perrault at Paris
—these offices stimulated invention ; perhaps the
memory of the Great Fire was enough to stir ingenious effort without their aid.

Now, there was a pearl-button maker named
Newsham, at Cloth Fair, not far distant from Pye
Corner, who obtained patents for improvements in fire-
engines in 1721, and again in 1725; while the *Daily
Journal* of April 7th, 1726, gives a report of one of
his engines which discharged water as high as the
grasshopper on the Royal Exchange. This apparently
was not only due to the great compression of air in
the air-chamber, but also to the peculiar shape he
gave to the nozzle of the jet ; and it is said he was
able to throw water to a height of a hundred and
thirty feet or more.

In France a man named Perier seems to have been
busy with fire-engines, though how far he worked
independently of others we cannot tell.

The hose and suction-pipe are said to have been invented by two men named Van der Hide, inspectors
of fire-extinguishing machines at Amsterdam about
1670. The hose was of leather, and enabled the water
to be discharged close to the fire. It is worthy of
note that this invention also appears to have been
after the Great Fire of London.

Remembering, therefore, that Newsham was probably indebted to others for the important air-chamber and flexible leathern hose—though how far he was indebted we cannot say—we must regard him as the Father of the Modern Fire-Engine in England. Especially so, as his improvements have been regarded as in advance of all others in their variety and value. It is also worthy of note that the first fire-engines in the United States were of his construction.

Little is known of Newsham's life. The reasons leading him, a maker of pearl buttons, to turn his attention to fire-engine improvement are not clear. At his death in 1743, the undertaking passed by bequest to his son. The son died about a year after his father, and the business then came into the hands of his wife and cousin George Ragg, also by bequest ; and the name of the firm became Newsham & Ragg.

One of Newsham's engines may be seen in the South Kensington Museum to-day, having been presented to that institution by the corporation of Dartmouth. The pump-barrels will be found to measure $4\frac{1}{2}$ inches in diameter, with a piston-stroke of $8\frac{1}{2}$ inches. The original instructions are still attached, and are protected by a piece of horn.

The general construction of Newsham's engines appears to have been something like this :

The body, which was long and narrow, measured about 9 feet by 3 feet broad ; this shape enabled it to be wheeled in narrow streets, and even through doorways. Along the lower part of the body, which was swung on wheels, ran a pipe of metal, which the water entered from a feed-pipe. The feed-pipe was intended to be connected with a source of supply ; but if this failed, a cistern, attached to the body of the engine, could be filled by buckets, while a strainer was

placed at the junction between the cistern and the
interior pipe to prevent dirt or gravel from entering it.
On the top of the body was built a superstructure,
which looked like a high box—greater in height than
in breadth, and larger at the top than at the bottom.
This box contained the all-important air-chamber and
the pumps. The water in the interior pipe was forced
into the air-chamber by the two pumps, and then
thrown on the fire through a pipe connected with a

EARLY MANUAL FIRE-ENGINE.

hose of leather projecting from the top of the air-
chamber. This pipe descended within the chamber
almost to the bottom, so that when water was pumped
into the air-chamber it flowed round the bottom of the
pipe, and prevented any ingress or egress of air. As
the water rose, the air already in the chamber became
compressed in the top part of the chamber, and in turn
exerted its power on the water.

The pumps were worked by levers, one on each
side of the engine, and alternately raised and

lowered by the men operating the machine ; while this manual-power was much increased by one or two men working treadles connected with the levers, and throwing the weight of the body on each treadle alternately.

The principle of the force-pump may be thus briefly explained :

When a tight-fitting piston working in a cylinder is drawn upward, the air in the cylinder is drawn up also, and a partial vacuum created ; if the cylinder is connected with water not too far distant by a pipe, the water will then rush upward to fill the vacuum. Then, if the bottom of the cylinder be fitted with a valve opening upward only, it is closed when the piston is pushed down again ; and the water would burst the cylinder, if enough power were applied to the piston, but escape is afforded along another pipe as an outlet, which in the case of the fire-engine opens into the air-chamber, and which is opened and closed by another valve. Thus is the water not only raised from the source of supply, but is forced along another channel.

And the modern fire-engine—which we date from Newsham's engines in England about 1726—is a combination of the principles of the force-pump and of the air-chamber, which acts by reason of the great elasticity of compressed air.

Other inventors made improvements as well as Newsham, namely, Dickenson, Bramah, Furst, Rowntree, and others, though the differences were chiefly in details. An engraving mentioned in an old work of reference sets forth that a London merchant named John Lofting was the patentee and inventor of the fire-engine. His invention must have been since the Great Fire, because the Monument is depicted in one

corner of the engraving and the Royal Exchange in
another. Rowntree made an engine for the Sun and
some other fire-offices, which protected the feed-pipe
more efficiently from mud and gravel ; and Bramah
devised a hemispherical perforated nozzle, which
distributed water in all directions, so that the ceilings,
sides, and floor of a room would become equally
drenched.

Bramah also applied the rotary principle to the
fire-engine. He studied the principles of hydraulics,
and introduced many improvements into machinery
for pumping, a rotary principle being one of them.
He attained this object by changing the form of
the cylinder and piston, the part acting directly on
the water being shaped as a " slider," and working
round a cavity in form of a cylinder, and maintained
in its place by a groove. He applied the rotative
principle to many objects, one being the fire-engine.
His fire-engine was patented in 1793 ; but we cannot
discover that it changed any vital principle of the
machine, which, as we have seen, consists in essence
of a movable force-pump, steadied and strengthened
by a compressed air-chamber and a flexible delivery-
hose.

Joseph Bramah, however, is doubtless best known
to fame as the inventor of the hydraulic press, though
he is also celebrated for the safety-lock which bears
his name. He was a farmer's son, and was born at
Stainborough in Yorkshire in 1748 ; but an accident
rendering him lame, he was apprenticed to a carpenter.
Engaging in business as a cabinet-maker in London,
he was employed one day to fit up some sanitary
appliances, and their imperfections led him to devise
improvements. He took out his first patent in 1778

and this contrivance proved to be the first of a long
series. His lock followed, and then, assisted in one
detail by Henry Maudslay, he introduced his hydraulic
press, a machine which he foresaw was capable of
immense development.

Several of his improvements are concerned with
water, such as contrivances connected with pumps
and fire-engines, and with building boilers for steam-
engines.   It is also said he was one of the first
proposers of the screw-propeller for steamships.
Altogether, he was the author of eighteen patents ;
though it has been pointed out that he improved and
applied the inventions of others, rather than originated
the whole thing himself.   While he contributed
improvements to the fire-engine, the vital principle
of the air-chamber and the flexible hose remained the
same. ¯Up to about the year 1832, the larger engines
generally in use in London seem to have thrown some
eighty-eight gallons a minute from fifty to seventy
feet high.

The next notable development was the application
of steam to work the force-pumps.   But this addition,
which was made about 1830 by John Braithwaite, also
did not alter the principle of the air-chamber.

John Braithwaite came of an engineering family.
He was born in 1797, the third son of John Braith-
waite, the constructor of one of the first diving-bells.
The ancestors of the Braithwaites had conducted an
engineer's business, or something analogous to it, at
St. Albans ever since the year 1695.

The younger John entered his father's business, and
from 1823, after his father and brother died, conducted
it alone.   Those were the days when steam was
coming into vogue, and he began to manufacture high-

pressure steam-engines. Together with Ericsson, he constructed the "Novelty," the locomotive which competed in the famous railway-engine contest at Rainhill in 1829, when Stephenson's "Rocket" won the prize. Braithwaite's engine, though it did not fulfil all the conditions of the competition, yet is said by some to have been the first locomotive to run a mile a minute—or rather more, for it is held to have covered a mile in fifty-six seconds. He used a bellows to fan the fire ; and in his steam fire-engine, he also employed bellows, though on one day of the Rainhill contest the failure of the bellows rendered the locomotive incapable of doing work.

In the fire-engine, the bellows were worked by the wheels of the machine, and eighteen or twenty minutes were required to raise the steam. At the present time, a hundred pounds of steam can be raised in five minutes in the biggest engine of the London Brigade, this result being due, in one respect at least, to the use of water-tube boilers.

Braithwaite's engine of 1830 was fitted with an upright boiler, and was of scarcely six horse-power ; but, nevertheless, it forced about fifteen gallons of water per minute from eighty to ninety feet high. The pistons for the steam and water respectively were on opposite ends of the same rod, that for steam being 7 inches in diameter, and for the water $6\frac{1}{2}$ inches, and both having a stroke of 16 inches.

The engine was successful in its day. During an hour's work, it would throw between thirty and forty tons of water on a fire ; while another engine, also made by Braithwaite, threw the larger quantity of ninety tons an hour.

The steam fire-engine was first used at the burning

of the Argyle Rooms in London in 1830 ; it was also used at the fire of the English Opera-House in the same year, and at the great fire at the Houses of Parliament in 1834. But, curiously enough, a great prejudice existed against it, and the engine was at length destroyed by a London mob. The fire-brigade were also against it. So Braithwaite gave it up ; but he built a few others, one at least being for Berlin, where it seems to have given great satisfaction.

Braithwaite, who became engineer-in-chief to the Eastern Counties Railway, also applied steam to a floating fire-engine, and constructed the machinery so that the power could be rapidly changed from propelling the vessel to operating the pumps.

The brigade could not long disregard the use of steam. In 1852, their manual-float was altered to a steamer, the alterations being made by Messrs. Shand & Mason. Six years later, the firm made a land steam fire-engine, which, however, was sent to St. Petersburg ; and then in 1860—thirty years after Braithwaite had introduced the machine—the London Brigade hired one for a year. The experiment was successful, and a steam fire-engine was purchased from the same makers. But only two steam fire-engines were at work at the great Tooley Street fire.

Then, in July, 1863, a steam fire-engine competition took place at the Crystal Palace, the trials lasting three days. Lord Sutherland was chairman, and Captain Shaw, who was then chief of the London Brigade, was honorary secretary of the competition committee. In the result, Merryweather & Son won the first prize in the large-class engine, and Shand & Mason the second prize. Shand & Mason also took the first prize in the small class, and Lee & Co. the second prize in the small class. The value of the steam fire-engine was fully established.

At the present time, Messrs. Shand & Mason have an engine capable of throwing a thousand gallons a minute ; while one of the water-floats of the London Brigade will throw thirteen hundred and fifty gallons a minute. These powerful machines form a striking development of Newsham's engine of 1726, and afford a remarkable contrast to the old fire-quenching appliances of former times.

But while the development of the modern fire-engine had been proceeding, a not less remarkable organization of firemen had been growing. It arose in a very singular, and yet under the circumstances a not unnatural, manner. And to this part of the story we must now turn our attention.

CHAPTER V.

EXTINGUISHMENT BY COMPANY. THE BEGINNINGS OF FIRE INSURANCE.

" CANNOT provision be made against loss by fire ? "

Looking at the terrible ruin caused in 1666, prudent men would naturally begin to ask this question. And some enterprising individual declared that a scheme must be launched whereby such provision might be made.

So, although proposals and probably attempts for fire insurance had been made before, by individuals or clubs, and by Anglo-Saxon guilds ; yet we read that " a combination of persons "—which, in the words of to-day, we suppose means a company—opened "the first regular office for insuring against loss by fire" in 1681.

Of course, another speedily followed. That is our English way. But both of these have disappeared. One, however,—the appropriately named Hand-in-Hand, which was opened in 1696, — still survives, and added life-insurance business in 1836. The Sun was projected in 1708 and started in 1710, the Union followed four years later, the Westminster in 1717, the London in 1720, and the Royal Exchange in the same year.

Therefore, the close of the seventeenth and the beginning of the eighteenth centuries saw the practice

LONDON FIREMAN IN 1696.

of fire insurance well established in Britain as an organized system. Now, these offices not only undertook to repay the insurers for losses, but also to extinguish the fires themselves. This latter, indeed, was fully regarded as an integral part of their business. Thus, one of the prospectuses of an early fire-office states that "watermen and other labourers are to be employed, at the charge of the undertakers, to assist at the quenching of fires." And it is worthy of note that, while the earliest men employed were watermen, the London Fire-Brigade to-day will only accept able-bodied sailors as their recruits.

The offices dressed their men in livery, and gave them badges ; the men dwelt in different parts of the city, and were expected to be ready when any fires occurred. Even to-day the interest of the companies

FIRE-INSURANCE BADGES.

in the extinguishment of fires is recognized, and their early connection therewith maintained ; for they pay the London County Council £30,000 annually toward the support of the brigade.

By the beginning of the nineteenth century, the fire-offices had notably increased in numbers. Thus, in 1810 there were sixteen, and some of their names will be recognized to-day. In additon to the Hand-in-Hand and the Sun, were the Phœnix (1782), the Royal Exchange, the North British (1809), the

4

Imperial (1803), and the Atlas, dating from 1808 ; there was also the Caledonian, dating from 1805.

Each company fixed its badge to the building insured, a course which appears to have been suggested by the Sun, and adopted so that the firemen of the different companies might know to which office the burning house belonged.

The badge was stamped in sheet-lead, and was painted and gilded ; but the badges for the firemen appear usually to have been of brass, and were fixed to the left arm. Each company not only kept its own engines and its staff of firemen, but also clad its men in distinctive uniforms. The dress for the Sun Office consisted of coat, waistcoat, and breeches of dark-blue cloth, adorned with shining brass buttons. The brass badge represented the usual conventional face of the sun, with the rays of light around, and the name placed above.

The helmet was of horse-hide, with cross-bars of metal. It was made of leather inside, but stuffed and quilted with wool. This quilting would, it was hoped, protect the head from falling stones or timbers, dangers which are still the greatest perils threatening firemen at their work.

By-and-by, Parliament made some effort towards organizing fire extinction. In 1774, a law was passed, providing that the parish overseers and churchwardens should maintain an engine to extinguish fires within their own boundaries. These engines were doubtless manned in many parishes, especially in rural districts, by voluntary workers, who sometimes were probably not even enrolled in an organized voluntary brigade ; the police also in certain places undertook fire duty. But " what is every one's business is no one's business," and for various reasons numbers of these parish fire-engines fell into disuse.

In short, the organization for the extinguishment of fires was thoroughly unsatisfactory. The men belonging to the different companies were too often rivals, when they should have been co-workers; each naturally gave special attention to the houses bearing their badges. We obtain a remarkable picture of the inefficiency prevailing in a letter from an eye-witness, Sir Patrick Walker, in No. 9 of the *Scots Magazine* in 1814. It refers to Edinburgh, but doubtless is true of other places.

Sir Patrick had taken an active part in endeavouring to

ROYAL EXCHANGE FIREMAN.
(From a portrait.)

arrest a conflagration, and he remarks on "a total absence of combined and connected aid, which must often render abortive all exertions." The chief defect, he declares, lies "in having company engines, which creates a degree of jealousy among the men who work them." When all success depended on their united efforts, then they were most discordant. There

were often more engines than water to adequately
supply them, consequently no engine had probably
enough to be efficient. The remedy, he held, was
to abolish all names or marks, and form the whole
into one body on military principles.

Curiously enough, the brigade that was formed in
London has come to be regulated rather on naval
than on military principles ; but the essence of Sir
Patrick's suggestion was undoubtedly sound. He also
complained greatly of the waste of water by hand-
carrying, which, moreover, created great confusion.

These grave defects were, no doubt, also felt keenly
by the London fire-offices, and in 1825 some of them
combined to form one brigade. They were the Sun,
the Phœnix, the Royal Exchange, the Union, and
the Atlas; and seven years later, in the memorable
year 1832, all the more important companies united.

In this action they were led by Mr. R. Bell Ford,
director of the Sun Fire-Office. The organization
then formed was called the London Fire-Engine
Establishment, and had nineteen stations and eighty
men. It was placed under the superintendence of
Mr. James Braidwood, a name never to be forgotten
in the story of fire-brigades and their work.

But to learn something of this great man and his
daring deeds and noble career, we must change the
scene to Edinburgh.

CHAPTER VI.

THE STORY OF JAMES BRAIDWOOD.

" Something must be done ! "

Many an Edinburgh citizen must have expressed this decision in the memorable year 1824. Several destructive fires had occurred, and at each catastrophe the need of efficient organization was terribly apparent. It seemed as though the whole city would be burned.

Then the police took action. The commissioners of the Edinburgh police appointed a committee, and a Fire-Engine Corps, as it was called, was established, on October 1st of the same year. The new organi-zation was to be supported by contributions from various companies, from the city of Edinburgh, and from the police funds.

" But who was to superintend it ? "

Now, a gentleman had become known to the com-missioners, perhaps through being already a superin-tendent of fire-engines ; and though only twenty-four years of age, he was appointed.

His name was James Braidwood. He was born in 1800 in Edinburgh, and was the son of a builder. Receiving his education at the High School, he after-wards followed his father's business. But in 1823, he was appointed superintendent of the fire-engines, perhaps owing to his knowledge of building and carpentry ; and when the corps was established, he was offered the command.,

He proceeded to form his brigade of picked men. He selected slaters, house-carpenters, plumbers, smiths, and masons. Slaters, he said afterwards, become good firemen ; not only from their cleverness in climbing and working on roofs—though he admitted these to be great advantages—but because he found them generally more handy, and ready than other classes of workmen.

They were allowed to follow their ordinary occupations daily ; but they were regularly trained and exercised every week, the time chosen being early in the morning. Method was imparted to their work. Instead of being permitted to throw the water wastefully on walls or windows where it might not reach the fire at once, they were taught to seek it out, and to direct the hose immediately upon it at its source.

This beneficial substitution of unity, method, skill, and intelligent control for scattered efforts, random attempts, lack of organization, and discord in the face of the enemy, was soon manifest.

Five years after the corps had been established under Mr. Braidwood, the *Edinburgh Mercury* wrote : " The whole system of operations has been changed. The public, however, do not see the same bustle, or hear the same noise, as formerly ; and hence they seem erroneously to conclude that there is nothing done. The fact is, the spectator sees the preparation for action made, but he sees no more. Where the strength of the men and the supply of water used to be wasted, by being thrown against windows, walls, and roofs, the firemen now seek out the spot where the danger lies, and, creeping on hands and feet into the chamber full of flame or smoke, often at the hazard of suffocation, discover the exact seat of danger, and, by bringing the water in contact with it, obtain immediate mastery over the powerful element with which they have to contend. In this daring and dangerous work, men have occasionally fainted from heat, or dropped down from want of respiration ; in which case, the next person at hand is always ready to assist his companion, and to release him from his service of danger."

Not only exercising great powers of skilful management, Braidwood showed remarkable determination and presence of mind in the face of danger. Hearing on one occasion that some gunpowder was stored in an ironmonger's shop, which was all aflame, he plunged in, and, at imminent risk of his life, carried out first one cask from the cellar, and then, re-entering, brought out another, thus preventing a terrible explosion.

In 1830, Mr. Braidwood issued a pamphlet dealing with the construction of fire-engines, the training of firemen, and the method of proceeding in cases of fire. In this work he declared he had not been able to find any work on fire-engines in the English language— a state of things which testifies to the lack of public interest or lack of information in the matter in those days. The book is technical, but useful to the expert before the era of steam fire-engines.

But in a volume, issued a few years after his death, Mr. Braidwood takes a comprehensive glance at the condition of fire extinguishment in different places. The date is not given ; but it was probably about 1840.

In substance he says : " On the Continent generally, the whole is managed by Government, and the firemen are placed under martial law, the inhabitants being compelled to work the engines. In London, the principal means . . . is a voluntary association of the Insurance Companies without legal authority ; the legal protection by parish engines being, with a few praiseworthy exceptions, a dead letter. In Liverpool, Manchester, and other towns, the extinction of fires by the pressure of water only, without the use of engines, is very much practised. In America, the firemen are generally volunteers enrolled by the local governments, and entitled to privileges."

From this bird's-eye view, it will be seen that organization for fire extinction and the use of efficient appliances for fighting the flames were still in a very unsatisfactory state ; yet the increasing employment of lucifer-matches and of gas in the earlier years of the nineteenth century tended to increase conflagrations.

Moreover, it is curious that the public seemed but little aroused to this unsatisfactory condition of affairs. Perhaps they saw their way to nothing better ; perhaps, if they took precautions, they regarded a fire as unlikely to occur in their own house, even if it might happen to their neighbour. Whatever the cause, they seem to have been but little stirred on the subject.

It was probably Mr. Braidwood's pamphlet of 1830 that led to his appointment as chief of the newly-formed London Fire-Engine Establishment. The publication showed him to be an authority on the subject, and one likely to succeed in the post. He came with the cordial good wishes of his Edinburgh friends. The firemen presented him ⸱with a gold watch, and the committee with a piece of plate.

He was ever careful of his men. He watched their movements, when they were likely to be placed in positions of peril ; and he would not allow any man to risk unnecessary danger. Yet he was himself as daring as he was skilful, and never shrank from encountering personal risk.

This was the sort of man who came to lead the

London Fire-Engine Establishment. He found it a small force, composed of groups of men accustomed formerly to act in rivalry, and having between thirty and forty engines, throwing about ninety gallons a minute to a height of between seventy and eighty feet, and also several smaller hand-hauled engines, comparatively useless at a large fire. In addition to the establishment of the associated companies, there were about three hundred parish engines and many maintained at places of business by private firms.

By his energy and skill, Mr. Braidwood kept the fires in check, and came to be regarded as a great authority on fire extinguishment and protection from fire. On these subjects, he was consulted in connection with the Royal Palaces and Government Offices, and held an appointment as a chief fire inspector of various palaces and public buildings. He became an Associate of the Institute of Civil Engineers, and read several papers before that body, and also before the Society of Arts, on the subject of the extinction and prevention of fires.

The force under his command was increased from eighty to a hundred and twenty men ; but it still remained the Establishment of the Fire-Offices. Throughout the country, the extinguishment of fire continued largely in the hands of voluntary workers, assisted by various authorities, even the fire-brigades being sometimes supplemented by the police and the water companies, as well as the general public.

And then an event occurred, which not only thrilled London with horror, but probably led to one of the most remarkable developments in the efforts for fire extinction that England had known.

CHAPTER VII.

THE THAMES ON FIRE.   THE DEATH OF BRAIDWOOD.

ABOUT half-past four o'clock in the afternoon of June 22nd, 1861, an alarm of fire reached the Watling Street station.

The firemen turned out to the call ; but little did they think, as they hurried along, that the fire to which they were summoned would burn for a whole month, and would become known as one of the most serious in the history of London.

The call came from Tooley Street, on the south side of London Bridge. Some jute in the upper part of a warehouse had been discovered smouldering, and bucketsful of water had been thrown upon it ; but the smoke became so thick and overwhelming, that the men were compelled to desist, and the flames grew rapidly.

By this time the alarm had been sent to Watling Street. Quickly the fire-engines arrived on the spot, and the men found dense masses of smoke pouring from buildings at Cotton's Wharf. A number of tall warehouses, rising up to six stories high, and filled with inflammable goods, stood here and near by, among the goods being oil, tallow, tar, cotton, salt-petre, bales of silk, and chests of tea. In spite of all efforts, the fire burned steadily on, and dense volumes of smoke poured forth.

Mr. Braidwood had speedily arrived, and two large floating-engines, in addition to others, were got to work. He stationed his men wisely, and huge jets of water were speedily playing on the fire.

Great excitement soon rose in the neighbourhood. Surging crowds of eager people thronged the streets approaching the wharf, and a dense assemblage pressed together on London Bridge. Even the thoroughfares on the opposite side were blocked. But the spectators

JAMES BRAIDWOOD.

could see little just then, except thick clouds of smoke and great jets of water. On the river, vessels struggled to escape from the proximity of the burning building ; while on land, the police forced back the people from the surrounding streets, so as to give greater freedom to the firemen.

Then, about an hour after the alarm had been given, a loud explosion startled the people ; a bright tongue of flame shot upward through the smoke, and seemed to strike downward also to the ground, while the whole building became a sheet of fire.

The neighbouring buildings became involved ; rivers of fire burst out of windows, ran down walls, and actually flowed along the streets. It even poured on to the waters of the Thames itself. Melted tallow and oil flowed along as they burned, like liquid fire. No wonder the conflagration spread rapidly. Less than two hours after the call had been received—that is, at about six o'clock—the fire had extended to eight large warehouses.

The heat now became overpowering. Drifting clouds of smoke obscured the calm evening sky, and spread like a pall overhead. In spite of all efforts, the fierce conflagration gained continually on the men ; it leaped over a space between the buildings, and attacked a block of warehouses on the opposite side. The roaring of the flames, the thick smoke, and the curious, disagreeable smells arising from the various goods which were burning, became almost unbearable.

The men suffered greatly from exhaustion; and Mr. Braidwood, seeing their distress, procured refreshments. He was dividing them among the men as he stood near the second building which had caught fire, when again a loud explosion rent the air, and the wall of the warehouse was seen to be falling.

" Run for your lives ! " was the cry ; and the men, seized for once with panic, rushed away. Mr. Braidwood and a gentleman with him followed; but unhappily they were not in time, and with a loud crash the huge wall fell upon them, and crushed them to the ground with tons of heavy masonry.

" Let us save them ! " cried the men ; and a score hurried to the spot. But again a third explosion occurred, a mass of burning material was hurled on the fatal heap, all around fell the fire, and rescue was seen to be hopeless.

As if in triumph, the flames swept on and mounted higher. Wharf after wharf was involved, and warehouse after warehouse. The Depôt Wharf, Chamberlain's Wharf, and others caught fire. Night seemed

THE TOOLEY STREET FIRE, 1861.

turned into day by the blaze. Ships near the wharves, laden with the same inflammable materials of oil, and tar, and tallow, became ignited ; and the blazing liquids poured out on the river, forming a lake of fire a quarter-mile long by a hundred yards wide.

People crowded everywhere to see the sight. They thronged house-tops and church-steeples. Boatmen ventured near to pick up such goods as they might be able to find, and were threatened with dire peril. Some fainted from the heat. A barge drifted near

with three men aboard, who were so overcome that
they could not manage their cumbersome craft ;
a skiff approached sufficiently near to rescue the
men, after which the barge drifted nearer still, and
was burnt.

Though greatly dispirited by the loss of their captain, the firemen fought doggedly on. But still their
efforts seemed unavailing. Flakes of fire fell in all
directions, and huge volumes of flame flashed upward
to the sky. The whole of Bermondsey seemed in
peril, and at one period the fire blazed for close upon
a quarter-mile along the river-bank.

Through the night more engines clattered up from
distant stations, and the firemen fought the flames
at every step of their destructive career. Tons of
water were poured upon each building as it became
threatened, only, however, to yield in course of time.

The wind saved the old church of St. Olave's, and
also London Bridge Station ; but the fire raged along
the wharves. Sometimes great warehouse walls fell
into the river with a gigantic splash, revealing the
inferno of white-hot fire raging behind them.

At length the fire reached Hay's Wharf, which was
supposed to be fireproof, and for long it justified the
name. But at last it also yielded ; the upper part
began to blaze, and, in spite of the quantities of water
thrown upon the roof and walls, the fire gradually
increased.

Now beyond the building lay a dock, in which were
berthed two ships. The tide had been too low to
allow of their removal. If they could not be towed
out in time, the fire would probably seize them, and
thus be wafted over the dock to the other side.

Would the tide rise in time to allow the ships to

be hauled out ?   It was a critical moment, and the
firemen must have worked their hardest to keep the
building from flaming too quickly.

Gradually the tide flowed higher and higher.   No
matter what happens in the mighty city, twice in the
day and night does the Thames silently ebb and flow ;
and now the quiet flowing of the tide helped to save
the great city on its bank.   Just in time two tugs
were able to enter the dock.   The towing-ropes were
thrown aboard ; but even as the vessels were passing
out, the flames, as if determined not to lose their prey,
darted from the building, and set the rigging of one
ship aflame.

But the firemen were as quick as their enemy.   An
engine threw a torrent of water on the burning ship,
and promptly quenched the flames.   And so, amid the
plaudits of the huge crowds on both sides of the river,
the two ships were slowly towed to a place of safety,
and the fierce fire was left face to face with the empty
dock.

The quiet dock was successful.   The wide space
filling up with water from the flowing tide stopped
the progress of the fire.   This stoppage must have
occurred about five o'clock on the following morning ;
but within the area already covered by the conflagration, fire continued to burn for a month.

Even after the first seven days, a fresh explosion
and flash of flame showed the danger of the conflagration, now fortunately confined within limits.   In
fact, July 22nd had dawned before it was entirely
extinguished, the total loss being estimated at about
two millions sterling.

Nearly all the goods destroyed were of the most
inflammable description.   There were nine thousand

casks of tallow and three hundred tuns of olive oil,
beside thousands of bales of cotton, two thousand
parcels of bacon, and other valuable merchandise.
The tallow, no doubt, burned the fiercest and the most
persistently. Melting with the intense heat, it poured
out into cellars and streets, where much of it speedily
caught fire. The floors of nine vaults, each measuring
100 by 20 feet, were covered two feet deep with
melted tallow and palm oil, and all helped to feed
the fire. No wonder it burned for days, if such
material fed the flames, although the firemen continued to pour water on the ruins. Some of the
tallow, found floating on the river, was collected, and
sold at twopence per pound.

Mr. Braidwood's body was found on June 24th, so
charred as to be scarcely recognizable. He was buried
at Abney Park Cemetery, and was accorded the honour
of a great public funeral. The London Rifle-Brigade
attended, as well as large bodies of firemen and of
the police, and an immense concourse of the general
public. So large a multitude, it was said, had not
attended any funeral since the obsequies of the Duke
of Wellington.

A proposition was made to raise a public fund for
the benefit of Mr. Braidwood's widow and six children,
and a large sum was subscribed ; but it was announced
that the Insurance Companies had amply provided for
his family.

The neighbourhood of Southwark, where the fatal
fire occurred, has been the scene of many remarkable
conflagrations. In the same year as the famous Tooley
Street fire, Davis's Wharf at Horselydown was burnt,
involving a loss of about £15,000 ; while at a large fire
at Dockhead two or three years later, vast quantities

of saltpetre, corn, jute, and flour were consumed. A brisk wind favoured the flames, and hundreds of tons of saltpetre flashed up into fire. Bright sparks and flame-coloured smoke floated over the conflagration, and were wafted by the wind, accompanied by deafening reports and great flashes of fire.

Numbers of other conflagrations have occurred in this neighbourhood. The streets were narrow, and the district was full of warehouses, containing all kinds of merchandise, which burnt like tinder when fairly ignited. Imagine coffee and cloves, sulphur and saltpetre, oil, turpentine, and tallow all afire ! What a commingling of odours and of strange-coloured flame !

The bacon frizzles ; the corn parches and chars ; the flour mixes with the water, then dries and smoulders in the great heat, and smells like burning bread ; the preserved tongues diffuse an offensive odour of burning flesh ; while the commingling of cinnamon and salt, mustard and macaroni, jams and figs and liquorice, unite to make a hideous combination of coloured flames, sickening smells, and thick and lurid smoke. The huge warehouses built in this district since the closing years of the eighteenth century are filled with all kinds of goods from various parts of the world ; but of all the disastrous fires which have ravaged the district, the great Tooley Street fire of 1861 has been the worst.

Moreover, it will always be memorable for the death of Braidwood. Even now you may hear men in the London Fire-Brigade speak of Braidwood or Braidwood's time, and his memory has become a noble tradition in the service. So great an authority had he become on the subject of fire extinction, and so highly

5

was he held in public esteem, that his terrible death in the performance of his duty was regarded as a national calamity.

But the conflagration also revealed with startling clearness the inadequacy of the Companies' Fire Establishment. More appliances and more men were wanted. The companies were asked, " Will you increase your organization ? " And their answer, put briefly, was, " No."

Thereupon, in 1862, a Parliamentary Commission was instituted to enquire into the matter, and in due time the commission reported. It recommended that a brigade should be established ; the companies consulted with the Home Secretary and the Metropolitan Board of Works ; and in 1865 an Act was passed placing the brigade under the Metropolitan Board, the change to take place as, and from January 1st, 1866.

This was practically the establishment of a Municipal Fire-Brigade, though it was also provided that every company insuring property for loss by fire in London should contribute to the cost of the brigade at the rate of £35 for every million pounds of the gross amounts insured, except by way of reassurance ; the Government were also to pay £10,000 a year for the protection of public buildings ; while the Metropolitan Board itself was empowered to levy a rate not exceeding a halfpenny in the pound in support of the organization.

In 1863, the Fire-Engine Establishment had increased to a hundred and thirty men with twenty stations ; but the Metropolitan Board were given power to construct further engines and stations, to act in conjunction with a salvage corps, to obtain the services of the men, and to divide the metropolis into suitable districts. Such powers would enable the Board greatly to strengthen the brigade.

The Act also provided that the firemen should be placed under command of an officer, to be called the Chief Officer of the Metropolitan Fire-Brigade ; and a gentleman was appointed who had had experience of similar duties at Belfast, and who was for long to be popularly known in London as Captain Shaw.

And on the very day when the new arrangements came in force a great fire occurred, as if to roughly remind the organization of its responsibilities and test its powers.

CHAPTER VIII.

A PERILOUS SITUATION. CAPTAIN SHAW. IMPROVEMENTS OF THE METROPOLITAN BOARD AND OF THE LONDON COUNTY COUNCIL.

" THE dock is on fire ! "

On New Year's Day, 1866, some hours after St. Katherine's Dock had been opened for work, several persons came running to the gates from the adjoining streets, crying loudly, " The dock is on fire ! "

At first the policemen would not believe the report. " We can see nothing," said they.

" But flames are bursting from the roof! Look! look ! "

And before long the policemen were convinced that a serious fire was, indeed, in progress. It was in the upper floors of a division of a block of warehouses

named F, six stories high, and by eleven o'clock they
were blazing fast.

" Fire ! Fire ! "

The alarming cry rang through the dock, and superintendents, dock managers, and policemen hurried to
the spot ; while gangs of dock labourers were taken
off their work, and set to quench the fire with buckets.

The conditions were somewhat similar to those of
the great Tooley Street fire of five years or so before.
The fire broke out on a floor where bales of jute and
coir fibre were stored ; and a huge heap of these
goods was seen to be burning, and sending forth such
a suffocating and blinding smoke, that the men were
compelled to retreat.

"Shut the iron doors ! " shouted the officers ; and
one after another the iron doors between the different
warehouses were closed, though with one exception.
This was the door connecting the fifth floor of F
Warehouse with the fifth floor of H Warehouse. It
was open wide, and one man after another endeavoured
to close it by crawling towards it on the floor. But
the smoke was so suffocating that the men had to
be dragged back almost unconscious before they could
reach the door.

Meantime, the dock fire-engines and hydrants had
been got to work, and the dock engineer was able to
turn on full pressure, so that soon powerful jets of
water were thrown on the flames. A hydrant is,
briefly, an elbow-shaped metal pipe, permanently fixed
to a main water-pipe ; and when the fireman attaches
his hose to it, he can get at once a stream of water
through the hose at about the same pressure as the
water in the main.

The flames were spreading furiously, and the two

upper floors of F Warehouse were blazing fast, throwing out such dense clouds of smoke, that the neighbourhood was darkened as by a thick fog. The block of warehouses on fire towered up six stories high, and occupied half of the northern side of the dock next to East Smithfield. They formed a huge pile about 440 feet long by about 140 feet deep, the import part of the dock lying on the south side with its ships.

The block was built in a number of divisions or bays, each measuring about 90 by 50 feet, and separated by strong walls, which rose from basement to roof. Happily, the communication between these divisions was afforded by double folding-doors of iron, a space of about three feet existing between the double doors ; they were believed to be fireproof ; and with the one exception they were closed.

But, like the Tooley Street buildings, these warehouses were chiefly stored with very combustible materials. Tallow was here, which played such a bad part in 1861 ; spirits were here also, palm oil, tons of dyewood, flax, jute, and cotton. Labourers had been at work for some hours when the alarm was given, and men were busy on every floor. They were receiving the goods from the quays, and wheeling them along through the building, when the fire was discovered.

And now Captain Shaw, the chief who succeeded Braidwood as the head of the fire-brigade, dashed up with a steamer from Watling Street, which was then the headquarters of the brigade. He had received the alarm at about twenty minutes to twelve o'clock, and had telegraphed to all subsidiary stations.

Captain Shaw, who afterwards became Sir Eyre

Massey Shaw, K.C.B., was born the same year as the steam fire-engine was first used—*viz.*, in 1830. He was the son of Mr. B. R. Shaw, of Monkstown, County Cork, and in due time entered the army. Retiring in 1860, he became chief of the Belfast Borough Forces, including police and fire-brigade, being appointed in the next year the chief of the London Fire-Brigade.

THE FIRST COMPLETE FLOATING STEAM FIRE-ENGINE, 1855.

Not only did he telegraph for land steam fire-engines to the conflagration ; but a large steam-float, usually kept off Southwark Bridge, was also quickly under way. Soon he had eight land steamers and from seventy to eighty men on the spot, while he himself directed in person.

Mr. Collett, one of the Dock Company's secretaries, worked hard, and often at great peril ; Mr. Graves and Mr. Stephens, also officials of the company, were busily engaged in directing removal of valuable materials ; while about seventy men employed by Cubitt & Co. in rebuilding a warehouse, destroyed by fire in the previous October, rendered assistance.

The little army found themselves face to face with a difficult task. The fire was now burning furiously, and the smoke was well-nigh overpowering. The flames had reached the fourth, fifth,

and sixth floors, and seemed working downward :
while the burning jute sent forth such dense volumes
of smoke, that the men were forced back again and
again. But bravely they returned to their task ; and
taking advantage of the moments when the clouds
cleared, they directed the hose to the most needful
points.

For six hours the fire raged, until all the three
upper floors were destroyed, and the third floor seri-
ously damaged. The scene in the waning winter afternoon was sufficiently striking as the smoke gradually
cleared and the blackened ruins became dimly visible.
They were very dangerous, for the walls appeared
likely to topple over at the slightest provocation.

About five o'clock, the firemen seemed to have
gained the mastery, and Captain Shaw returned home ;
but later in the evening he was summoned again.
Most mysteriously the flames had burst forth once
more in fresh places, the upper parts of two adjacent
warehouses of the same block had caught, and were
in flames. By eleven o'clock the fire was blazing as
furiously as ever.

Captain Shaw returned with new relays of men to
assist those on the spot ; and during the night and
all the next day the force was busily at work. On
the Monday night two firemen were so overcome by
the smoke that they had to be removed, being nearly
suffocated ; but happily they recovered, and no life
was lost during the fire. The streams of melted grease
flowed from the burning warehouse into the quay, and
thence to the dock basin, where by-and-by they cooled
and solidified, looking something like snow on a
frozen lake. Thirteen steam fire-engines and one float
continued to throw immense quantities of water on

the burning building ; but the fire was not really subdued until the morning of January 3rd.

A few engines remained on the Wednesday and the Thursday, and threw water on the heated ruins, to cool them down and quench any latent fire ; while on January 4th, men were busy skimming the dock basin, —which was thickly covered with the solid tallow and oil,—and loading the mass into barges.

After the conflagration, engines were employed in pumping water out of the vaults where it had collected, and as much jute was found injured by water as destroyed by fire. No doubt, it was the jute and the tallow and oil which rendered the conflagration so obstinate ; but it was also found that while water collected to a great extent in some parts, yet it did not penetrate to other parts of even the same floor— a result which, perhaps, was due to the method of packing the jute.

In the end, about three-parts of the block of ware-houses was burned. The amount of tallow in the four burning buildings was calculated to range between two and three thousand casks, some of which appear to have been saved ; but several hundred barrels of cocoanut oil and palm oil were lost as well, and the coir fibre, flax, and jute burnt reached to a very large quantity, the total pecuniary loss being estimated at over £200,000.

This great fire proved a terrible object-lesson. For about two days and nights the engines and appliances of the brigade, with some two-thirds of the men, were engaged at this one conflagration. What if another great fire had broken out in those dark January days ? The situation was fraught with the gravest peril.

No doubt, voluntary aid at fires used often to be relied upon, and in 1861 payment was given to assistants. But the Metropolitan Board now had the means of strengthening the brigade, and they proceeded

SIR EYRE M. SHAW, K.C.B.

to use it. In marked contrast to the 130 men and 20 stations of the Fire Establishment of 1863, were the 591 firemen and 55 land fire-engine stations of the brigade in 1889, when it passed over to the London County Council—figures which show a notable development.

Further, there were also 83 coachmen and pilots,
131 horses, 150 engines (55 being worked by steam),
155 fire-escapes, and other ladders, with 33 miles of
hose.   By this time (1889) many provincial towns
had established a fire-brigade on the London
plan.

The London County Council, having no restriction
as to powers of rating, adopted Captain Shaw's
recommendations—made in April, 1889—of a large
increase in the brigade, and resolved to add 138
firemen, 4 new stations, with steamers and manuals,
and 50 fire-escapes, and to raise the number of
electrical fire-alarms to over 600.

Since then, the increase has still continued, until in
1898 the brigade had an authorized fire-staff of nearly
1,100 men, with a certain number of store-keepers, etc. ;
while the telegraphic arrangements and distribution
of stations were rendered so complete, that 100 men
could be concentrated within fifteen minutes at any
dangerous area for large fires.

Furthermore, out of the authorized staff, 134 men
are on watch by day, and 369 at night, giving a total
of 503 constantly on duty during the twenty-four
hours—a force that compares wonderfully with the
total strength of about 130 men at Braidwood's death
in 1861.

This brigade strength of 1,048 included about 80
officers, 824 firemen, 96 coachmen, 17 pilots, and 32
men under instruction.   To these must be added
seventeen licensed watermen for navigating tug-boats,
river-engines, etc., and also stores and office clerks.
But twenty-four additional firemen, however, have
been sanctioned, so that the complete staff would
reach to about 1,080 men—a remarkable development

of the staff of 80 men of the London Fire-Engine Establishment of 1832.

These figures are only given to show how greatly the brigade has grown ; for in the course of a few years, it is not improbable that the numbers may be still further increased.

The number of stations has also been remarkably augmented. The 19 stations of 1832 have grown into nearly 200 for divers uses. Thus, there are 189 fire-escape stations, 59 stations with engines, 57 with hose-carts, 9 with hose- and ladder-trucks, 16 permanently established in centres of wide streets with fire-extinguishing and life-saving appliances, and 4 river stations.

The appliances of the brigade have also greatly increased. There are 230 fire-escapes and police-ladders, 59 land steam fire-engines, 57 six-inch manuals, 7 small manuals called curricles, 175 horses which we may rank as most useful appliances, and 24,284 hydrants.

These last-named are very important. They not only afford a ready and efficient means of throwing water on conflagrations, a means which is fast rendering the manual-engines of less and less importance ; but they also yield a quick and ready method of water supply. Thus, in the year 1897 there were only three cases of unsatisfactory water supply.

In addition to 24,284 hydrants of the London County Council, the corporation of the City have 800 hydrants, which are used for watering the streets as well as for extinguishing fires. In the year 1897, no fewer than 466 fires were put out by hydrants and stand-pipes.

The increase of hydrants has been very conspicuous

under the County Council. Thus, in March, 1889,
the number was but 8,881, showing that no fewer
than 15,403 were added during the first eight years
of the Council's existence. No doubt, still more will
follow. On March 31st, 1898, hydrants had been
fixed or ordered in $97\frac{1}{2}$ square miles of the county
area, leaving a comparatively small space unprovided
with these appliances. This space will doubtless be
shortly supplied, and it is not unreasonable to suppose

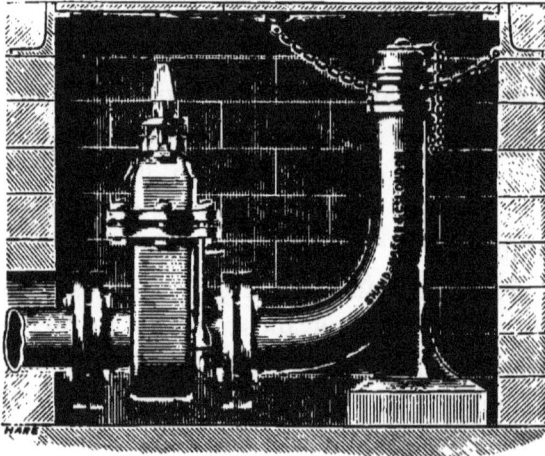

FIRE-HYDRANT PLACED UNDER THE PAVEMENT.

that, with the 800 in the City, the metropolis will ere
long be sown with a total of about 30,000 hydrants,
which, as the twentieth century dawns, may be regarded as among the most effectual means of fighting
the fire at the disposal of the brigade.

The establishment of these excellent appliances
dates from 1871, and is bound up with the system
of constant water supply. By the Metropolis Water
Act of that year, it was provided that a water company,
after giving a constant supply, must notify the fact

to the local authority—now the County Council—
which must then specify the fire-plugs or hydrants
required, and the Council has the power under the
Act of requiring water companies to provide a constant supply within parts of their districts. Hydrants
are fully charged from the main, and have a commanding cock or tap attached, so that a supply of
water can be obtained at once.

The use of these appliances is very important.
Planted at convenient and commanding spots,—often
at the corners of streets or roadways, and at varying
distances apart, ranging from fifty to about four
hundred feet, according to the circumstances of the
locality, and marked also, not only by the plate in the
pavement, but by the letter H, placed in a conspicuous
position near,—the fireman can now, at almost a
moment's notice, find the hydrant, and obtain an ample
supply of water for his engine, or even a jet of water
for the fire, before an engine is on the spot. Very
different from the troublesome and hindering work
of floundering about, possibly in fog or rain or snow,
to find the fire-plug, and then to find the turncock
which governed the plug. On snowy or foggy nights,
the difficulty and delay were sometimes very great ;
and the substitution of an extensive system of hydrants,
with their quickly-obtained water-jets for the old
fire-plugs, may rank as one of the most efficient means
of fire extinction in the closing years of the nineteenth
century.

Firemen being thus interested in the pressure of
water in the mains, an apparatus for recording the
pressure automatically was fixed up at the fire-brigade
headquarters at Southwark Bridge Road in November,
1898. A clock stands at the top of the instrument,

and under the clock is a roll of paper, having the hours of day and night marked upon it, and divided into sections. A small pipe connected with the main runs under the big engine-room, and acts upon mechanism beneath the paper roll, and the clock and the column of water, and its pressure per inch, are marked in red ink upon the sheet, varying perhaps from forty up to seventy-five or even eighty pounds per square inch.

At noon each day the sheet can be removed, and forms a permanent record of the variation in water pressure in the mains of the neighbourhood.

But if the number of hydrants is large, the area to be protected by the brigade is also very large. Including the ancient city of London, which is estimated to cover about a square mile, the area measures about 118 square miles. Of these, twelve are estimated by the fire-brigade committee to be covered by parks and open spaces, where fire-hydrants will probably never be needed. This leaves, however, a net area of 106 square miles, extending from Sydenham to Highgate, and from Plumstead to Roehampton, to be efficiently protected by the brigade.

Another means of water supply has been suggested. In his evidence at the Cripplegate Fire Enquiry, Mr. John F. Dane, an ex-officer of the Metropolitan Fire-Brigade, suggested that at the centre of the junction of the most important streets surrounded by large buildings underground tanks should be placed, and supplied by the main water-pipes. The tanks would be empty until required, and would be under the control of the brigade, while the hydrants should still be maintained for service. Such tanks were in use at Leeds and at Salford.

The objection is raised, however, that the streets of the City are already too crowded with pipes, while advantage of the pressure from the water-main is lost, and also the vacuum caused by the engine.

Noticing other improvements, we observe that the number of fire-alarm posts has also been greatly increased. The alarm consists of a red post in the street, with a glass face at the top front. The glass is readily broken, and the handle within it pulled, when a loud electric bell rings at the nearest fire-station. The Post-Office provides and maintains the fire-alarms ; and Commander Wells, chief officer of the brigade, has devised a portable telephone, which can be plugged into a fire-alarm post, and a message sent by it from a fire to the station. Arrangements have been made with the Post-Office to supply the telephones and make the plug-holes. Over 2,380 fire-alarms were raised in 1897, of which 363 were maliciously-given false alarms. Practical jokes of this kind have been heavily punished, as they richly deserve.

Many false alarms are also given which cannot be regarded as malicious, but are genuine mistakes, such as of supposed chimney fires. Over 500 of these were recorded in one year. In 1898, the number of malicious false alarms was happily less—*viz.*, 270 ; while the full record of false alarms reached 830.

The total number of fires in the metropolis in that year was 3,585—an average of nearly ten per day. This total gives an increase of 571 above the average ; but only 205 out of the whole 3,585 were serious. There seems no doubt but that the public are learning to use the fire-alarms more readily and to give earlier intimation of fires. But, as the chief officer points

out, while everybody knows the nearest letter-box,
very few comparatively even now seem to know the
nearest fire-alarm. Lamp-posts near the alarms are
now painted red, and are fitted with a red pane of
glass in order to attract attention ; and we imagine
the probability is that the alarms will be increasingly
used at even the slightest appearance of fire.

Not only is each fire-station connected with a dozen
or more fire-alarms in its neighbourhood, but it is
also in electric communication with other fire-stations.
There are 114 lines of telephone between the stations,
and sixteen between brigade- and police-stations ;
while electric communication exists between stations
and ninety-eight public or other buildings. In fact,
the whole fire-brigade establishment is bound together
by a web of electric wire, the centre being the headquarters at Southwark.

The remarkable organization of the brigade, famous
for its leaders, famous for the bravery and skill of its
men, and famous for the number and variety of its
efficient appliances, has been a growth of comparatively
few years. Starting in 1825 with the union of a few
fire-office companies, it grew in seventy-three years
to a remarkably strong and increasing force, with a
multitude of hydrants, stations, horsed escapes, fire-
alarms, and other appliances.

The development attained in these seventy-odd years,
as compared with the hundreds of years before, is
surely marvellous, though doubtless some seeds of
the development—as in the introduction of the modern
fire-engine—were sown before. But step by step it
has proceeded, utilizing now the discoveries of science
and now the work of the engineer, until it has reached
its great position of usefulness and of high esteem.

It would be tedious to mark every detail of development. The work begun by his predecessors was carried still further by Captain Shaw, and under him

COMMANDER WELLS.

the London Brigade became one of the most efficient in the world.

He retired with a well-deserved pension in 1891, after about thirty years of service, and was succeeded by Mr. J. Sexton Simonds. Five years later Mr.

6

Simonds retired ; and in November, 1896, Commander Lionel Wells, R.N., was appointed chief officer. The brigade has also a second officer—Mr. Sidney G. Gamble ; and in January, 1899, a third officer was appointed—Lieutenant Sampson Sladen, R.N.

A few months after his accession, and in answer to the request of the fire-brigade committee of the County Council, the chief officer submitted a scheme for additional protection, including certain regulations of brigade management.

Of this scheme, the more prominent features were the introduction of horsed fire-escapes, and the distribution of the men in small stations, with horses, whence they can be speedily concentrated wherever required. In short, the chief officer's object is that, at any call, the firemen may be able, if the machine leave the station at once, to arrive at the fire within five minutes' time ; while the principle of station-work should be that each station is responsible for a certain area in its neighbourhood.

The committee agreed with the opinions of the chief officer, and on February 8th, 1898, the full Council adopted the committee's proposals. Steps were forthwith taken to carry out the scheme, which thus marks another stage of development.

But let us visit the headquarters, and see for ourselves something of this great organization actually at work.

CHAPTER IX.

A VISIT TO HEADQUARTERS.

" WE light our fires differently from everybody else," says the foreman. " We put shavings on top, the wood next, and the coal at the bottom ; then we strike a steam-match, and drop it down the funnel, and, behold ! the thing is done." It was the engine fire of which the foreman spoke, and he was pointing to one of the magnificent steam fire-engines at the head-quarters of the London Brigade.

HEADQUARTERS, METROPOLITAN FIRE-BRIGADE, SOUTHWARK.

" Here is a steam-match," he continued, " kept in readiness on the engine.   It is like a very large fusee, and is specially made for us.   Water won't put it out."

He strikes the match, and it burns with a large flame.   He plunges it into some water near by, and

it still continues to burn.   It evidently means to
flame until the engine fire is burning fast.

The wood also is carefully prepared, being fine deal
ends, specially cut to the required size; while the
coal is Welsh—the best for engine-boilers.

These details may seem trivial ; but they assist in
the rapid kindling of the
engine fire, which is not
trivial.   But the rapid
kindling of the fire is not
the only reason why the
brigade raises steam so
quickly in its engines ; in
addition, a gas-jet is always
kept burning by the boiler,
and maintains the water
at nearly boiling-point before the fire is lighted.
This was a method adopted
by Captain Shaw.   But
even this arrangement does
not explain everything.

To fully understand the
mystery, we must leave
this smart engine, shining
in scarlet and flaming
with brass, and go upstairs
to the instruction-room for recruits.

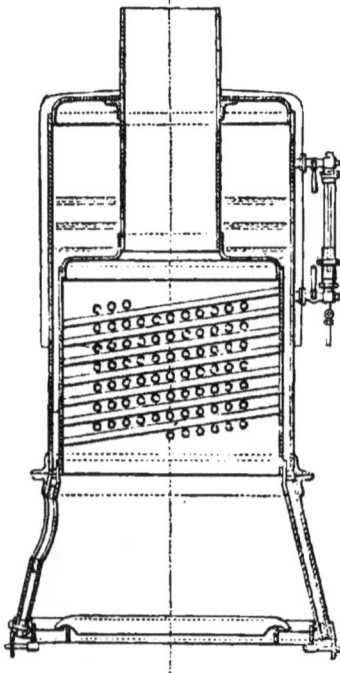

SECTION OF A STEAM
FIRE-ENGINE BOILER.

Here we can see a section of the engine fire-box
and boiler.   It is very interesting and very ingenious.
But probably a novice would ask, " Where is the
boiler ?   I see little else but tubes."

That is the explanation.   The tubes chiefly form
the boiler; for they are full of water, and they com-

municate with a narrow space, or "jacket," also full
of water, and which reaches all round the fire-box.
This fire-box is held in a hollow below the tubes,
which are placed in rows, one row across the other,
just at the bottom of the funnel and above the fire-
box. When, therefore, the flaming steam-match is
dropped down the funnel, it finds its way straight
down between the crossed mass of tubes to the
shavings beneath ; and the tubes full of the hot water
are at once wrapped in heat from the newly-kindled
and rapidly-burning fire. Every particle of heat and
smoke and flame that rises must pass upward between
the tubes. Furthermore, the hot water rises and the
colder falls, so that there is a constant circulation
maintained. The colder water is continually descending to the hottest tubes ; and when bubbles of steam
are formed, they rise with the hot water to the top.
A space is reserved above the tubes, and around the
funnel, called the " steam-space " or " steam-chest,"
where the steam can be stored ; the steam pressure at
which the engine frequently works being a hundred
and twenty pounds to the square inch.

The result of all these ingenious arrangements is
that, starting with very hot water, a hundred pounds
of steam can be raised in five minutes.

" But," it may be asked, " why is a fire not always
kept burning, and steam constantly at high pressure?"

The answer is that a constant fire, whether of coal
or of oil, would cause soot or smoke to accumulate ;
while the Bunsen gas-burner affords as clear a heat
as any, and maintains the water at a great heat, or
even at boiling-point.

Near the funnel, but not so high, rises a large,
gleaming metal cylinder, closed and dome-shaped.

This is the indispensable air-chamber, without which
even the powerful force-pumps could not yield so steady
and persistent a stream.

A small air-chamber is now added to the suction-
pipe by which the water is drawn to the engine.
The use of the air-chamber in connection with this
pipe greatly steadies the engine, the vibration caused
by the throbbing of the powerful machinery as it
draws and forces along such a quantity of water
being very great. The nozzle of the hose belonging
to one of the largest steam fire-engines measures
$1\frac{1}{4}$ inch in diameter, some nozzles being as small
as $\frac{3}{4}$ inch; and a large column of water is being
constantly driven along the hose at a pressure of a
hundred and ten pounds to the square inch, and
forced through the narrow nozzle; here it spurts
out, in a large and powerful stream, to a distance
of over a hundred feet. It is obvious, therefore,
that the power exerted by the steam-driven force-
pumps and air-chamber is very high; and although
such an engine may be in some folks' opinion only
a force-pump, it is a force-pump of a very elaborate
character; and not inexpensive, the average price
being about £1,000.

Every steam fire-engine carries with it five hundred
feet of hose. The hose is made in lengths of a
hundred feet, costing about £7 a piece, without the
connections. If you examine a length, you will find
it made of stout canvas, and lined with india-rubber,
the result being that, while it is very strong, it is yet
very light.

Miles of it are used in the service; and upstairs in
the hose-room you will find a large stock kept in
reserve. Every piece is tested before being accepted.

POWERFUL STEAM FIRE-ENGINE FOR THE METROPOLITAN FIRE-BRIGADE.

Capacity, 350-400 gallons per minute.   Delivered to the brigade, February 9th, 1899, by Messrs. Shand, Mason, & Co.

Water is forced through it by hydraulic power at a pressure of three hundred pounds to the square inch, so that when at work, with water rushing through at a hundred and ten pounds' pressure, it is not likely to split and spill the liquid on the ground. The splitting of hose in the face of a fierce fire would be a great calamity. When charged with water, its weight is very heavy ; and to enable it to be carried more easily, a loop called a " becket " is attached at distances of about ten feet.

The greatest care is taken of the hose. When it is brought back, drenched and dripping, from a fire, it is cleaned and scrubbed, and then suspended in the hose-well to dry.

The hose-well is a high space, like a glorified chimney-shaft, without the soot, where the great lengths of canvas pipe can be hung up to dry. They are, in fact, not used again until they are once more in the pink of perfection. The outside public see the fire-brigade and their appliances smartly at work at big fires, but little know of the numerous details of drill and of management which are instrumental in producing the brilliant and efficient service.

Look, for another instance, at the manuals' wheels. You will find them fitted with broad, wavy-shaped iron tyres, which extend over the side of the wheel and prevent it from tripping or slipping over tramway-lines in the headlong rush through the streets. And should a horse fall as he is tearing to the fire, that swivel-bar, which you will find at the end of the harness-pole, can be quickly turned, and in a moment the fallen steed is unhooked and helped to his feet again.

The horses are harnessed quite as quickly. Behind the engine-room and across a narrow yard you will

find five pairs of horses, and, like the men, some
are always on the watch. Here they stand, ready
harnessed, their faces turned round, and looking over
the strip of yard to the engines. The harness is
light, but efficient ; and the animal's neck is relieved
from the weight of the collar, as it is suspended from
the roof.

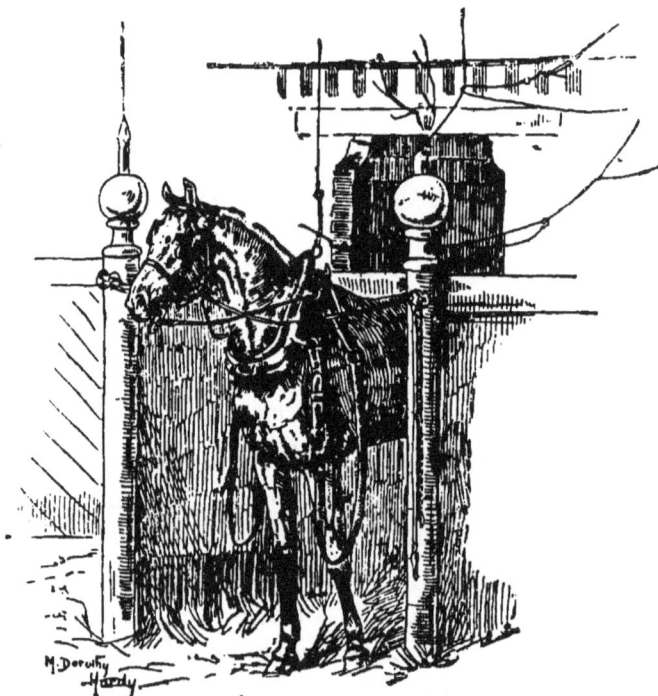

IN THE STABLES READY FOR ACTION.

Directly the fire-alarm clangs, the rope barring egress
from the stall is unswivelled, the suspender of the
collar swept aside, and the horse, eager, excited, and
impatiently pawing the ground, is led across the
narrow strip of yard, hooked on to the engine, and
is ready for his headlong rush through the streets.
Horses stand thus ready harnessed at all stations

where they may be kept ; and when their watch is over, they are relieved by others, even though they may not have been called out to a fire. So intelligent have some of these animals become, that they have been wont to trot out themselves, and take their places by the engine-pole without human guidance ; and so expert are the men and so docile the horses, that the whole operation of harnessing to the engine occupies less than a minute, sometimes, indeed, only about fifteen or twenty seconds. Every man knows exactly what to do, and has his place fixed on the engine. There is consequently no confusion and no overlapping of work.

A steam fire-engine has a " crew "—as the brigade call it—of one officer, one coachman, and four firemen. The officer No. 1 stands on the " near side " of the engine by the brake ; No. 2 stands on the other side by the brake ; No. 3 stands behind the officer, and No. 4 behind No. 2 ; No. 5 attends to the steam, and rides in the rear for that purpose ; while the coachman handles the reins on the box.

The positions are taken in a twinkling, the shed-doors open as swiftly, and away rush the impatient steeds, while the loud and exciting cry of " Fire! Fire!" rings from the firemen's throats as they speed along. Wonderfully that cry clears the way through the crowded streets. When the men arrive at the scene of action, the preparations proceed in the same orderly manner. Nos. 1 and 2 brake the wheels, and proceed to the fire ; while the coachman, if necessary, removes the horses, and is prepared to take back any message with them, No. 1 charging No. 2 to convey the message to the coachman. By the chief officer's plan, however,— whereby a portable telephone, carried on a fire-engine, can be plugged into a fire-alarm post,—a message can be sent back from a fire by telephone instead of by a coachman. Meanwhile, No. 3 is opening the engine tool-box, and passing out the hydrant-shaft, hose, etc. ; and No. 4 receives the hose, and connects it up with the water-mains, and places the dam or tank in which water is gathered from the hydrant. No. 3 is then busy with the delivery-hose, which is to pour the water on the flames ; and No. 5 connects the suction-pipe. When ready, No. 4 hurries away with the "branch," as the delivery-pipe with nozzle is called ; No. 3 helping with the hose attached to it—until sufficient is paid out—and connecting the lengths as required. Then, when all is finished, every one except the steam-man is ready to proceed to the fire, unless otherwise instructed. Every engine, it may be added, carries a turncock's bar, useful for raising the cover from the hydrants.

So each one has his recognized duties in preparing the apparatus, all of which duties are duly set forth in the neat and concise little pocket drill-book prepared by Commander Wells. The most complete organization must be in operation, otherwise a force of a hundred or a hundred and fifty men, no matter how brave and zealous, gathered at one fire would only be too likely to get in one another's way. And in a similar manner the crews of manual-engines and horsed escapes have all their duties assigned in preparing the machines.

During a conflagration, the superintendent of the district in which the fire occurs controls the operations under the superior officers ; for London is divided, for fire purposes, into five districts, which are known to the brigade by letters. A District is the West End,

and the superintendent's station is at Manchester
Square ; B District is the Central, and the superintendent's station is at Clerkenwell ; C District is
the East and North-East, with district superintendent's
station at Whitechapel : all of these three being
north of the Thames. The D District is the South-
East of London, with superintendent's station at New
Cross ; and the E District in the South-West, with

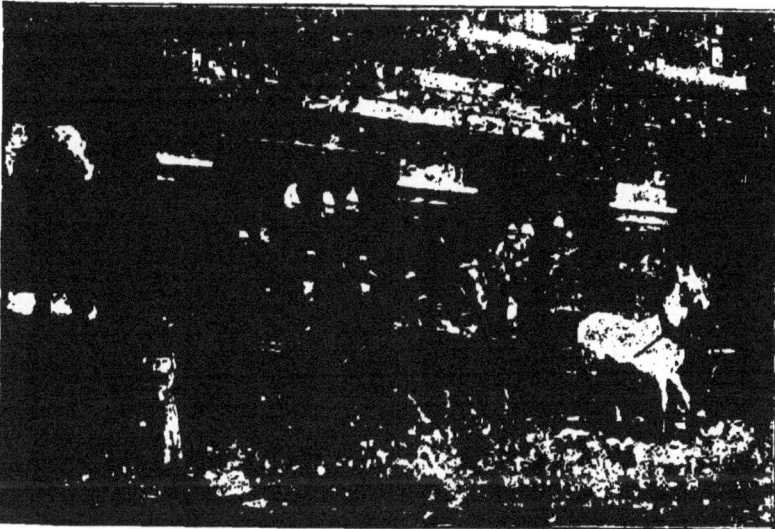

A TURN OUT FROM HEADQUARTERS AT SOUTHWARK.

superintendent's station at Kennington. The headquarters, which are known as No. 1, and which used
to be at Watling Street in the City, now occupy a
central position in Southwark Bridge Road, and
thence the chief officer can readily reach the scene of
a fire.

All these stations are in electric communication,
and all telegraph their doings to No. 1. The lines
stretch from No. 1 to the five district superintendents'
stations ; from there they extend to the ordinary

stations in each district ; and from these stations again
they reach to points such as street stations, and even
in some cases to hose-cart stations. The consequence
is, that superintendents and superior officers can
speedily arrive on the spot ; and that, if necessary,
a very large force can be concentrated at a serious
outbreak in a short time.

Thus headquarters knows exactly how the men are
all engaged, and the character of the fire to which
they may be called.   Electric bells seem always
clanging.   Messages come clicking in as to the
progress of extinguishing fires, or notifying fresh
calls, or announcing the stoppage of a conflagration.
And should an alarm clang at night, all the other
bells are set a-ringing, so that no one can mistake
what's afoot.

A list is compiled at headquarters of all these
fires, the period of each list ranging from 6 a.m.
to the same hour on the next morning. This list,
with such details as can be supplied, is printed
at once, and copies are in every insurance-office by
about ten o'clock.   The lists form, as it were, the
log-book of the brigade.   Some days the calls run
up to seventeen or more, including false alarms ;
on other days they sink to a far fewer number ;
the average working out in 1898 to nearly ten calls
daily.

The Log also shows the causes of fires, so far as
can be ascertained ; and the upsetting of paraffin-
lamps bulks largely as a frequent cause. The over-
heating of flues and the airing of linen also play
their destructive part as causes of fires. The airing
of linen is, indeed, an old offender. Evelyn writes
in his Diary, under date January 19th, 1686 : " This

night was burnt to the ground my Lord Montague's palace in Bloomsbury, than which for painting and furniture there was nothing more glorious in England. This happened by the negligence of a servant airing, as they call it, some of the goods by the fire in a moist season; indeed, so wet and mild a winter had

THE CHIEF'S OFFICE AT SOUTHWARK, METROPOLITAN FIRE-BRIGADE.

scarce been seen in man's memory." And now, more than two hundred years later, the same cause is prevalent.

But the upsetting and exploding of lamps is now, perhaps, the chief cause, especially for small fires; and more deaths occur at small fires than at large. This is not surprising, when we remember that such lamps

are generally used in sitting or bedrooms, where
persons might quickly be wrapped in flames or over-
whelmed with smoke.

Smoke, indeed, forms a great danger with which
firemen themselves have to contend. At a fire in
Agar Street, Strand, in November, 1892, a fireman was
killed primarily through smoke. He was standing
on a fire-escape, when a dense cloud burst forth and
overpowered him. He lost his grasp, and, falling
forty feet to the earth below, injured his head so
severely that he died.

Again, several men nearly lost their lives through
smoke at a fire about the same time at the London
Docks. The firemen were in the building, when thick
smoke, pouring up from some burning sacks, nearly
choked them. Ever ready of resource, the men quickly
used some hose they had with them as life-lines, and
slipped from the windows by means of the hose to
the ground below.

Nevertheless, dense smoke is not the greatest danger
with which firemen are threatened. Their greatest
peril comes from falling girders and walls, from
tottering pieces of masonry, and burning fragments
of buildings, shattered and shaken by the fierce
heat. Helmets may be seen in the museum at headquarters showing fearful blows and deep indentations
from falling fragments of masonry, and firemen would
probably tell you that they suffer more from this
cause than any other.

For small fires in rooms, little hand-pumps, kept
in hose-carts, are most useful. They can be speedily
brought to bear directly on the flames and prevent
them from spreading. These little pumps can be
taken anywhere ; they are used with a bucket, which

is kept full of water by assistants, who pour water into it from other buckets.

The fire, large or small, being extinguished, a message to that effect is sent to headquarters, and the firemen return, with the possible exception of one or two men to keep guard against a renewed outbreak. In the case of larger fires, perhaps half a dozen men and an engine will remain ; while on returning, the various appliances have all to be prepared in readiness to answer another alarm. It sometimes happens that a fireman may be on duty for many hours at a stretch, or may only have time to snatch an hour's sleep with clothes and boots on ; for nearly every hour a fresh alarm comes clanging into the station, telling of a new fire in some part of busy London. And for any real need, there is, I trow, no grumbling or complaint from the brave men. But the miscreant detected in raising a malicious false alarm would have scant mercy. He would be promptly handed over to the police, and the magistrate would punish him severely—perhaps with a month's imprisonment.

When not actually engaged at fires, the men find plenty to do in painting and repairing appliances, attending to horses, and keeping up everything to the pink of perfection. The hours on duty and for specified work are all marked -down in the brigade-station routine, general work commencing at 7 a.m., and ending at one, while allowing for a "stand easy" of fifteen minutes at eleven. The testing of all fire-alarms once in every twenty-four hours, excepting Sundays and before six o'clock at night, also forms part of the brigade-station routine. Every fireman, how-ever, has a spell of twenty-four hours entirely off duty in the fortnight ; but at all other times he is ready to

be called away.   Indeed, men on leave are liable to
be summoned in case of urgent necessity ; but such
time is made up to them afterwards.

Now, before being drafted into the effective ranks,
all the men have to pass through a three months'
daily drill at headquarters.   The buildings are very extensive, affording accommodation for about a hundred
men, thirty-five or so being the recruits.   In the
centre, enclosed by the buildings, stretches a large
square, in which the drill takes place.   To see the
combined drill is something like seeing the brigade
actually at work ; and this being Wednesday afternoon,
and three o'clock striking, here come the squad of
men marching steadily into the yard.

The evolutions are about to begin.

CHAPTER X.

HOW RECRUITS ARE TRAINED.

TRAMP, tramp, tramp !   Two lines of wiry, muscular
young men march into the centre of the yard.

" Halt ! Right about face ! "

Quick as thought the men pause and wheel around.

Indian clubs and dumb-bells !

The opening of the drill this afternoon is a course
of exercises with these familiar appliances ; but they
soon give place to other evolutions, such as jumping
in the sheet, practise with the engines, rescue by the
fire-escape, and the chair-knot.

Round and round whirl the clubs.   Every day some
section of the drill is taken ; but on Wednesday

afternoons, the whole or combined drill is practised. All candidates must have been sailors ; no one need apply who has not been at least four years an A.B. Further, they must be between the ages of twenty-one and thirty, and able to pull over the escape ; that is, they must be able to pull up a fire-escape ladder from the ground by the levers. The height of the ladder is about 28 feet, and the pull is equal to a weight of about 244 pounds. It is a hard pull, and a severe test of a man's strength ; but after the first

TESTS OF STRENGTH FOR MEN ENTERING THE FIRE-BRIGADE : PULLING UP THE ESCAPE.

twelve feet, the weight seems lessened, as the man's own weight assists him. In this test, as in some other things, it is the first step that costs. Should the candidate pass this test successfully, he is examined by the doctor ; finally, he comes to headquarters for his probationary drills.

" Open order ! "

The men break off from their gymnastic exercises, and in obedience to instructions some of them run for large canvas sheets, and spread them out, partly folded

on the ground. Then others calmly lie themselves
down on these sheets. What is going to happen ?

The recruits approach the recumbent figures, which
lie there quite still, and apparently heavy as lead ;
the lifeless feet are placed close together, and the
limp, inanimate arms arranged beside the body. Then,
at a word or a sign, the bodies are picked up as easily
as though they were tiny children, and carried over
the recruits' shoulders—each recruit with his man—
some distance along the yard. The men are practising
the art of taking up an unconscious person, overcome
may be by smoke, or heat and flame, and carrying
him in the most efficient manner possible out of
danger.

There is more in this exercise than might at first
appear. It might seem a comparatively easy task—
if only you had sufficient strength—to throw a man
over your shoulder and carry him thus, even leaving
one of your hands and arms quite free ; but you
would find it not so easy in the midst of blinding
flame and choking smoke ; you would find it not so
easy to pick your uncertain way through a burning
building and over flaming floors, over a sloping roof
or shaky parapet, and even down a fire-escape.

Hence the urgent necessity that the fireman should
be so well practised, that in a moment he can catch
up an insensible, or even conscious person in exactly
the most efficient manner, and, with hand and arm
free, be able to find his way quickly out of the fire.

He must be cool and clear-headed, dexterous, and
sure-footed, ready of resource, and quick yet reliable
in all his movements ; and to these ends, as to others,
the drill is directed.

Captain Shaw's advice to those beginning " in the

business of extinguishing fires" may be quoted here from his volume on " Fire and Fire-Brigades." " Go slowly," he says, " avoid enthusiasm, watch and study, labour and learn, flinch from no risk in the line of duty, and be liberal and just to fellow-workers of every grade."

But shouts of laughter are rising, as presently two or three of the recruits at the drill appear in a long flowing skirt, and look awkward enough in their unaccustomed garments as they stride along. They imitate women for the nonce, and are rescued in a similar manner, the men also carrying apparently lifeless figures down the ladders of the escapes.

The sheets, however, are used for other purposes of drill. See ! A group of men are opening one out, and carrying it below an open window some twenty-five feet above the ground.

There are fourteen or so of these men, and they grip the sheet firmly all round, and spread

ESCAPE-DRILL.

it out a little less than breast-high. A man appears at the window, twenty-five feet or so above. He is about to jump into the sheet far below.

At the cry he leaps, or rather drops, down plump into the sheet ; and the force of the fall is so great, that, unless these men were all leaning well backward, it would drag them toward the ground, and the rescued man sustain injury. As it is, they are all

dragged pretty well forward by the impact of the fall.

A person jumping like this into a sheet should drop down into it, not spring, as though intending to cover a great space. And the persons holding the sheet should lean as far backward as possible. If they simply held the sheet, standing upright in the ordinary way, no matter how firm the grip, they would probably all be dragged to the ground in a heap.

The jumping-sheet is made of the best strong canvas about 9 feet square, and strengthened with strips of webbing fastened diagonally across. The sheet is also bound round at the edges with strong bolt-rope, and is furnished with about a score of hand-beckets, or loops. If at a fire all other means of rescue be unavailable, the sheet should be brought into use. Volunteers, if necessary, should be pressed into the service, and instructed to stretch out the sheet by the beckets, holding it about two feet or so from the ground. They should grasp the becket firmly with both hands, the arms being stretched at full length, their feet planted well forward, but their heads and bodies thrown as far back as possible. Even then the volunteers will probably find great difficulty in maintaining the sheet, and preventing it from dashing on the ground. If possible, a mattress or pile of straw or some soft object should be placed on the ground beneath the sheet. The uninitiated have no idea of the weight of a body suddenly falling or jumping on to the sheet from a great height, and this occasion is one for the putting forth of all the strength of body and determination of will of which a man may be capable.

But, now the sheet is being folded, and men are

appearing on the roofs of the buildings above. A new exercise is beginning. Rescue by rope is now to be practised, and long threads of rope begin to appear. Imagine yourself a fireman on the top of a burning house, with smoke and flame belching out of the windows below, and agonizing screams for help ringing in your ears. No fire-escape is near, or, if near, not available ; for it sometimes happens that persons cannot be rescued by ladders, and the staircase is a mass of flames. What would you do ?

It is then that the firemen use the chair-knot, or, speaking popularly, they try rescue by rope. Every engine carries excellent rope of tanned manilla, and the fireman carries a rope about his body. Quickly the ends of the rope are fastened to two points, one on either side of the window—to a chimney-stack, if possible ; then, as sailors know how, by means of what is called a "tomfool's knot," loops and knots are made in the rope—one loop to be slung under the arms, and the other to support the knees, and together forming a sort of chair. Speedily the loops are adjusted round the person to be rescued, and then he is gradually lowered to the ground. A guiding-rope has been attached, and thrown to the men below, and is used by them to steady the person's descent, to prevent him from bobbing hither and thither, or to draw him out of reach of the flame and smoke.

This exercise being over, there is a rattle and a clatter, and into the yard dashes a horsed fire-escape. The men pounce upon it at once, and in a trice whip it off its carriage and wheel it to the building. The present escapes are great improvements on the old forms, and two men can extend it with ease.

The first or main ladder of the escape reaches about

24 feet high ; and in the 1897 pattern the 40-feet
ladders having one extension.   Other escapes have
extending-ladders rising to a height of 50 feet, and
even 70 feet, these being in three lengths.   But an
Act of Parliament now provides that all buildings
above a certain height must have means of exit
attached ; this generally takes the form of iron
ladders or stairways outside the building.   All parts
of an escape are as far as possible interchangeable,
and the ladder-vans are designed to carry any ladders
in the brigade.

And now the escape-drill is about to commence.
The machine is placed against the building, which we
must suppose to be burning.   Up runs a fireman, with
hands and feet on the rungs, to the window where the
top of the ladder rests.   If the window will not open
readily, he may, in case of real need, smash it with his
axe to obtain ready entrance.

Then, if you watched him closely, you would see he
did something which you would never think of doing.
He fastens the end of his rope to the rung of his
ladder, and, with the rest of the rope coiled over
his arm, disappears into the room.   The rope easily
runs out as he moves, and affords him a means of
speedily finding his way back to the window through
the smoke ; a very valuable arrangement it may
prove to be, when the fireman finds an insensible
person or a couple of children to rescue.

One child he carries in his arm, and the other he
throws across his shoulder, in the recognized brigade
manner ; and loaded thus, he gropes his way, guided
by his one free hand, along the rope.

Or there may be more than one adult to save.   Then
the rescued person is carried over the shoulder to the

top of the trough, or shoot of netting, with which some escapes used to be fitted at the back of the escape-ladder, and is slipped down it feet first to the firemen waiting below; while the plucky fireman above returns for the next person in peril.

The fireman will probably follow the last down the shoot by turning a somersault and coming down head first; meantime, holding the other's hands, and regulating the speed of the descent by pressing his knees and elbows against the sides of the netting. But without the shoot he descends by the ladder.

Should the fire occur at a house surrounded by garden-wall, shrubs, or forecourt, the machine is wheeled as close as possible, and the extension or additional ladders can be placed at a somewhat different angle from the first, so as to bridge over the intervening space and reach the farthest window. The ladders of fire-escapes may also be useful substitutes for water-towers. A water-tower is a huge pipe, running up beside the ladder, or tower; and as three or four steamers play into the base of the huge pipe, the water is forced up it, and the jet at the top can then be directed anywhere into the burning building.

" But we don't want any water-towers," exclaimed a fireman ; " we can make one ourselves, if we need one." That is, by using the fire-escape ladders to obtain points of vantage.

We soon see this accomplished. With a rush of horses and a whiz of steam, a fire-engine tears into the yard, the steam raising the safety-valve at a pressure of a hundred and twenty pounds to the square inch.

Off leap the men, as though actually at a fire ; each

one attends to his prescribed duty ; and ere long you
see one of the men hurrying up the escape-ladder
bearing the branch in his hand—*i.e.*, the heavy nozzle
end of the hose. In a second the engine whistles,
there is a spurt of water, and the fireman directs the
jet from the distant head of the ladder to a tank in
the centre of the yard.

The beckets on the hose, placed at intervals of
seventeen and then twenty feet, over a hundred-feet
length, are made of leather ; and are most useful for
fastening it to a chimney or any point of vantage by
means of the fireman's rope. The weight of a hundred-
feet length when complete ranges from sixty to sixty-
five pounds, and when full of water much more.

The hose for the London Brigade is woven seamless,
of the best flax ; and the interior india-rubber lining
is afterwards introduced, and fastened by an adhesive
solution. Unlined hose is used by some provincial
brigades ; and it is contended that the water passing
through it keeps it wet, and therefore not liable to be
burned by the great heat of the conflagration. On the
other hand, the leakage is said to be a very objection-
able defect. The internal diameter of the hose is two
inches clear at the couplings, but a little larger within.

The steam-man is taught to remember the great
power he rules ; otherwise he may, by neglecting to
give the warning whistle, endanger his brother-fire-
man's life by suddenly sending the water rushing
through the hose, or bringing a great strain upon it,
when the men controlling it are not prepared.

It may appear an easy thing to stand on a ladder
or a house-top, and direct the jet on the fire ; but it is
not so easy to carry and to guide the long, heavy, and
to some extent sinuous pipe, full of the heavy water

throbbing and gushing through it at such tremendous pressure, especially when your foothold is none too secure.

A fireman lost his life one night, when holding the hose on the parapet of a roof in the Greenwich Road. He overbalanced himself, and fell crashing, head downward, sixty feet or more below, and met a terrible death.

Whether this fearful accident was entirely due to the heavy hose, we cannot say; but unless hose be laid straight, it is apt to struggle like a living thing. The reason is obvious. The water rushes through it at great pressure; and if the hose be not quite straight, the pressure on the bent part of the hose is so great that it struggles to straighten itself. Consequently, a fireman turning a stream will probably have to use a great deal of strength.

The increase in velocity of the water by the use of a branch and nozzle is, of course, very great. A branch-pipe is defined by Commander Wells as "the guiding-pipe from hose to nozzle." Some branches are made of metal; but leather branches are being substituted for long metal pipes. Some of these latter measured from 4 to 6 feet long, and were not only very cumbersome to carry, but often impracticable to use with efficiency inside buildings.

Leather branch-pipes are sometimes longer, and are tapered from 2 inches in diameter to 1½ inch at the nozzle. When, therefore, a stream of water from two to two and a half inches in diameter, forced along at a great pressure, and distending the hose to its utmost capacity, is driven through the narrowing path of the branch-pipe, it spurts out from the nozzle at a much higher velocity; and it is just this narrowing part of the hose which the fireman has to handle, and whence he directs the jet.

Some nozzles are like rose watering-can pipes, and are furnished with a hundred holes to distribute the water. These nozzles are useful in interior conflagrations and smoky rooms.

Yet, all important as is the engine-drill, and in-valuable as are the engines for serious conflagrations, it is interesting to read in the Brigade Report that in 1897 no fewer than 808 fires were extinguished by buckets, and 460 by hand-pumps, while 98 were extinguished by engines, and, as we have said, 466 by hydrants and stand-pipes.

The brigade bucket carried on the engine holds about 2½ gallons, and is made of canvas ; it is collapsible, cane hoops being used for the top and bottom rings. Drill is maintained even for bucket and hand-pump; and the latter appliance is so portable, that the whole of the gear pertaining to it, including two ten-feet lengths of hose, is carried in a canvas bag.

Hand-pumps are often used for chimney fires. Two men usually attend, and expect to find a bucket in the house. They pour small quantities of water on the fire in the grate, and allow as large a quantity of steam as possible to pass up the flue. When the fire in the grate is quenched, the men use the hand-pump on the fire in the lower part of the chimney, and then, mounting to the roof, pour water down the chimney.

As sometimes the ends of wooden joists are built into the flues, an examination should be made to discover if the lead on the roof or in any place shows signs of unusual heat, and the joists have caught fire ; for outbreaks of fire have been known to occur from this obscure cause. A comparatively simple but effective means of dealing with a chimney fire is to block up both ends of the chimney with thoroughly wet mats or sacks; while one of the easiest methods is to throw common salt on the fire. The heat decomposes the salt, and sets free chlorine gas—common salt being chloride of sodium, and chlorine being a gas which very feebly supports combustion, and tends to choke and dull a fire, if not to extinguish it entirely.

And so the drill goes on, with scaling-ladders and long ladders, hose-carts and horsed escapes, steamers and manual-engines, the object of the whole being, not alone to perfect the men in their knowledge of the gear and machines, and skill in using them, but also to develop quickness of eye, and readiness and firmness of hand. A systematic routine is followed by fully-qualified instructors, part of the course being theoretical and part practical; while about the year 1898 a new syllabus of instruction came into use.

Among other alterations, it was arranged that a selected officer should take charge of the recruits' drill for about two years, instead of engineers appointed at comparatively short intervals. Further, it was decided to permanently increase the authorized number of recruits, with the anticipation that never fewer than thirty men will be under instruction ; and to prohibit them, if possible, from being called away to engage in cleaning or other work, so that their instruction drill should never be interrupted.

When the men have passed through a three months' course of instruction, they should be ready to be drafted into the ranks as fourth-class firemen. The men in the brigade are divided into four classes ; in addition to which, there are coachmen, and licensed watermen for the river-craft, also engineers, foremen, and super-intendents, the whole being in charge of a chief officer and a second and third officer.

First aid to the injured is also included in the instruction of the men ; and the Recruits Instruction-Room and Museum contains a beautifully-jointed skeleton, kept respectfully in a case, for anatomical lessons.

Further, if you search the indispensable boxes on the engines, you will find among the mattocks and shovels,

RELICS OF THE BRAVE.

the saws and spanners and turncock's tools, a few medical and surgical appliances. Every engine carries a pint of Carron oil, which is excellent for burns. Carron oil is so called from the Carron Ironworks, where it has long been used, and consists of equal parts of linseed oil and limewater ; olive oil may be used, if linseed oil be not procurable. Carron oil may be used on rags or lint ; and triangular and roller bandages are carried with the oil, also a packet of surgeon's lint and a packet of cotton-wool. Accidents which are at all serious are, of course, taken as soon as possible to the hospital. But, alas ! some accidents occur which no

Carron oil can soothe, or hospital heal ; and on that roll of honour in the little room beside the big engine-shed, and in the blackened bits of clothing and discoloured, dented helmets in the museum in the instruction-room, you find ample demonstration that a fireman's life is often full of considerable risk.

These are the mute but touching memorials of the men who have died in the service ; to each one belongs some heroic tale.   Let us hear a few of these stories ; let us endeavour to make these charred memorials speak, and tell us something of the brave deeds and thrilling tragedies connected with their silent but eloquent presence here.

Listen, then, to some stories of the brigade.

CHAPTER XI.

SOME STORIES OF THE BRIGADE.

HERE are two tarnished and dented helmets of brass. They belonged respectively to Assistant-Officer Ashford and to Fourth-class Fireman Berg, who both lost their lives at the same great conflagration.

About one o'clock in the early morning of December 7th, 1882, the West London policemen, stepping quietly on their beat about Leicester Square, discovered that the Alhambra Theatre was on fire.

A fireman on watch within the building had made the same discovery, and with his comrade was working to subdue the flames.   But they proved too strong for the men.

The nearest brigade station was speedily aroused,

the news telegraphed to others, and ere long several
fire-engines had hurried to the spot. Quickly they
were placed at different points about the building,
and streams of water were thrown on the fire. But
in spite of all efforts, it gained rapidly on the large
structure.

The position was fairly high and central, and the
flames and ruddy glow in the sky were visible in
all parts of London ; even at that hour spectators
rushed in numbers to the scene and crowded the
surrounding streets. It was with difficulty that the
police could prevent them from forcing themselves
into even dangerous situations.

The heat was intense, and as far off as the other side
of the spacious square it struck unpleasantly to the
face. The flames darted high in the air as if in
triumph, and the huge rolling clouds of smoke became
illumined by the brilliant light. Several notable
buildings in the neighbourhood stood out clearly in
the vivid glow as though in the splendour of a
gorgeous sunset, while high amid the towering flames
stood the picturesque Oriental minarets of the building
as though determined not to yield.

The firemen endured a fearful time. Some stood in
the windows, surrounded, it seemed, by sparks of fire.
Mounting fire-escapes also, they poured water from
these points of vantage into the burning building. By
half-past one twenty-four steam fire-engines were at
work, and at that time the brigade had only thirty-
five effective steamers in the force. At about two
o'clock the minarets and the roof fell in with a
tremendous crash, and still the flames shot upward
from the basement.

Crash now succeeded crash ; girders, boxes, galleries,

all fell in the general ruin. Moreover, the fire leaped out of the building, and began to attack other houses at the back. A number of small and crowded tenements existed here, and the danger of an extended and disastrous fire became very great. But the efforts of the firemen were happily successful in preventing its increase to any considerable extent.

It was while working on an escape-ladder that Berg met with his death. An escape had been placed against the building next to the front of the theatre, and he was engaged in directing the jet of water from the extended or " fly " ladder fifty feet high, when from some cause—probably the slipperiness of the ladder-rungs—he lost his footing, and crashed head-foremost to the ground.

When taken up, he was found to be insensible ; and while the fearful flames were still raging, and his comrades were still at work, he was conveyed to the Charing Cross Hospital. Among other injuries which he had received was a fracture of the head ; and after lingering a few days, and lapsing into long fits of unconsciousness, he died.

Not long after Berg was admitted to the hospital on that fearful night, another fireman was carried thither from the same place. This sufferer was Assistant-Officer Ashford, who arrived at the fire in charge of an engine from Southwark. He was standing behind the stage, when a wall fell upon him and crushed him to the ground. His comrades hurried to rescue him, and he was quickly taken to the hospital ; but his back was found to be broken, and he had also sustained serious internal injuries. After lingering for a few hours in great pain, he died. He had been thirteen years in the brigade, and was married.

8

Several other accidents occurred at this great fire. At the same time that Ashford was stricken down, Engineer Chatterton, who was standing near him, was stunned, and narrowly escaped with his life. Four other firemen were also injured, one suffering from burns, one from sprain and contusions of the legs, one from falling through a skylight and cutting his hands, and one from slipping from a steam fire-engine on returning to Rotherhithe and breaking his arm. These incidents show how various are the heavy risks the firemen run in the course of their work.

When any member of the brigade dies in the execution of his duty, it has been customary to accord the body a public funeral, and Ashford's obsequies proved a very solemn and imposing ceremony. At eleven o'clock on December 14th, a large crowd assembled in Southwark Bridge Road, and detachments of officers and men had been drawn from various fire-stations, until nearly three hundred representatives of the brigade were present. A large number of policemen also joined the procession. It had a long way to traverse to Highgate Cemetery, where the burial took place. The coffin, of polished oak, was carried on a manual-engine, and covered by a Union Jack, the helmet of the deceased and a beautiful wreath subscribed for by members of the brigade being placed upon the flag. Three police bands preceded the coffin, and after it came mourning-coaches with the relatives of the deceased. Captain Shaw followed, leading, with Mr. Sexton Simonds, the second officer and the chairman of the brigade committee of the Board of Works ; then came the large body of firemen with their flashing brass helmets ; superintendents and engineers were also present, and the large contingent of police. Finally, followed six manual-engines in their vivid scarlet, and representatives of the salvage corps and of volunteer brigades. The procession marched slowly and solemnly, the bands playing the Dead March in "Saul." And thus, with simple yet effective ceremony, the crushed and broken body was borne through London streets to its last resting-place.

It may be interesting to trace here the chief particulars of the fire, to illustrate the working of the brigade. Of the firemen watching on the premises, one had gone his round, when about one o'clock, on going on the stage, he saw the balcony ablaze. He aroused Hutchings, another fireman who slept at the theatre, and the two got a hydrant to work, there having been several fitted in the building; they also despatched a messenger to Chandos Street station, which is quite near. The fire proved too strong for the hydrant to quench it; and when the manual-engine from the station arrived, a fairly fierce fire was in progress.

Meantime, directly the alarm had been received at Chandos Street, it was, as is customary, sent on to the station of the superintendent of the district, and thence it was circulated to all the stations in the district, and also to headquarters. Captain Shaw was soon on the spot, and directed the operations in person. Of course, such a call as "The Alhambra Theatre alight!" would cause a number of engines to assemble; and in truth, they hurried from all points of the district: they came from Holloway and Islington, from St. Luke's and Holborn. But soon "more aid" was telegraphed for; and then engines came flying from Westminster and Brompton, from Kensington and Paddington,

even from Mile End and Shadwell in the far east,
and from Rotherhithe, Deptford, and Greenwich across
the Thames. In rapid succession, they thundered
along the midnight streets, waking sleepers in their
warm beds, and paused not until the excited horses
were pulled up before the furious fire.

In fact, just within half an hour of the first call
at Chandos Street station, twenty-four steamers were
at work on the fire, and throwing water upon the
flames from every possible point. Captain Shaw was
assisted by his lieutenant, Mr. Sexton Simonds, and
Superintendents Gatehouse and Palmer. The contents
of the building were so inflammable, or the fire had
obtained such a firm hold, that the enormous quantities
of water thrown upon it appeared to exercise little
or no effect. But at length, when the roof had fallen,
the firemen seemed to gain somewhat on their enemy;
and they turned their attention to the dwellings in
Castle Street, and prevented the flames from spreading
there. Finally, three hours after the outbreak, that
is, about four in the morning, the fire was practically
suppressed. Several of the surrounding buildings
were damaged by fire and heat, and by smoke and
water.

In the dim wintry dawn, the scene that slowly
became revealed presented a remarkable spectacle.
Looking at it from the stage door, the blackened front
wall could be seen still standing, though the windows
had gone, and within yawned a huge pit of ruin.
Scorched remains of boxes and galleries, dressing-
rooms and roof, all were here ; while huge girders could
be seen twisted and rent and distorted into all manner
of curious shapes, which spoke more eloquently than
words of the fearful heat which had been raging.

The value of strong iron doors, however, was demonstrated ; for the paint-room had been shut off by these doors from the rest of the building, and the flames had not entered it.

But to turn to other relics in the museum. Here lies a terrible little collection,—a part of a tunic, a belt-buckle, an iron spanner, part of a blackened helmet, and part of a branch-pipe and nozzle. They are the memorials of a man who was burnt at his post.

Early in the afternoon of September 13th, 1889, an alarm was sent to the Wandsworth High Street fire-station. The upper part of a very high building in Bell Lane, occupied by Burroughs & Wellcome, manufacturing chemists, was found to be on fire. The time was then about a quarter-past two, and very speedily a manual-engine from the High Street station was on the spot.

A stand-pipe was at once utilized, and Engineer Howard, with two third-class firemen, named respectively Jacobs and Ashby, took the hose up the staircase to reach the flames. Unfortunately, the stairs were at the other end of the building, and the men had to go back along the upper floor to arrive at the point where the fire was burning.

Having placed his two men, Engineer Howard went for further assistance. Amid suffocating smoke, Jacobs and Ashby stood at their post, turning the water on the fire ; and their efforts appeared likely to be successful, when suddenly, a great outburst of flame occurred behind them, cutting off their escape by the staircase.

It was a terrible position,—fire before and behind, and no escape but the window !

Both men rushed to a casement, and cried aloud,

"Throw up a line!" The crowd below saw the men tearing at the window-bars and endeavouring to break them, while the fire rapidly spread towards them.

Could no help be given? Howard had endeavoured to rejoin the two men, and, finding this impracticable, turned to obtain external aid. The ladders on the engine were fixed together, but they fell far short of the high window. A builder's ladder was added; but even this extension would not reach the two men caged up high above in such fearful peril.

A moment or two of dreadful suspense, and then the crowd burst forth into loud cheers. Ashby was seen to be forcing his way through the iron bars. He was small in stature, and his size was in his favour. By some means, perhaps scarcely known to himself, he dropped down to the top of the ladder and clung there, and finally, though very much burned, he reached the ground in safety.

But the other? Alas! his case was far different. It is supposed that the smoke overcame him, and that he fell on his face; but he was never seen alive again. Engines rattled up from all parts of London, and quantities of water were thrown on the flames, but to no effect so far as he was concerned. When the fire was subdued, and the men hastily made their way to the upper floor, they found only his charred remains. He had died at his post, the smoke suffocation, it may be hoped, rendering him insensible to pain.

But an even more terrible accident happened to a fireman named Ford, in October, 1871. His death, after saving six persons, remains one of the most terrible in the annals of the brigade.

About two in the morning of October 7th, 1871, an

alarm of fire reached the Holborn station. The call
came from Gray's Inn Road; and Ford, who had

FIREMAN FORD AT THE GRAY'S INN ROAD FIRE.

charge of the fire-escape, was soon at the scene of
action. He found a fire raging in the house of a

chemist at No. 98 in the road, and the inmates were
crying for help at the windows.

Placing the escape against the building, he hurried
to a window in one of the upper floors, and, assisted
by a policeman, brought down five of the inhabitants
in safety. Still there was one remaining, and frantic
cries from a woman in a window above led him to
rush up the escape once more. He had taken her from
the building, and was conveying her down the escape,
when a burst of flame belched out from the first floor
and kindled the canvas "shoot" of the escape. In
a second, both the fireman and the rescued woman
were surrounded by fire.

Unable to hold her any longer, he dropped her to
the ground, where she alighted without suffering any
serious injury. But the fireman became entangled in
the wire netting of the machine, and it held him there
in its cruel grasp, in spite of all his struggles, while
the fierce fire roasted him alive.

At length, by a desperate effort, he broke the netting,
apparently by straining the rungs of the ladder; but
he himself fell to the ground so heavily, that his
helmet was quite doubled up, and its brasswork hurt
his head severely. His clothes were burning as he
lay on the pavement; but, happily, they were soon
extinguished, and he was removed, suffering great
agony, to the Royal Free Hospital in the Gray's Inn
Road. He lingered until eight o'clock on the evening
of the same day, when he died.

He was only about thirty years of age, and had
been four years in the brigade, where he bore a good
character. A subscription was raised for his widow
and two children, and his funeral was an imposing
and solemn ceremony. The coffin was borne on a fire-

engine drawn by four horses to Abney Park Cemetery, and was followed by detachments of firemen and of police.

It is a peculiarly sad feature of this case that, after saving so many lives, he should himself have succumbed, and that the very machine intended to save life should have been the cause of his death. At the inquest the jury added to their verdict the remark that, had the canvas been non-inflammable (means having been discovered to render fabrics non-inflammable), and had the machine been covered with wire gauze instead of the netting, Ford's life might have been saved. Considerable improvements have been made in fire-escapes since then, and machines of various patterns are in use in the brigade ; but, speaking generally, it may be said that the shoot, when used, is made of copper netting, which is, of course, non-inflammable.

Happily, all the brave deeds of the firemen do not meet with personal disaster. One brilliant summer afternoon in July, 1897, the Duke and Duchess of York were present at the annual review of the brigade on Clapham Common, and the Duchess pinned the silver medal for bravery on the breast of Third-class Fireman Arthur Whaley, and the good service medal was given to many members of the brigade. Whaley had saved two little boys from a burning building, and his silver medal is a highly-prized and honourable memorial of his gallant deed.

About one o'clock on the early morning of April 26th, 1897, a passer-by noticed that a coffee-house in Caledonian Road, North London, was on fire. Several policemen hurried to the spot ; but in three minutes from the first discovery the place was in flames. The

house was full of people. Mr. Bray, the occupier, was
apparently the first inmate to notice the fire from
within, and the others were soon aroused. The
terrified people appeared at the windows, and, impelled
by the cruel fire, threw themselves one after the
other into the street below. They numbered Mr. and
Mrs. Bray and four daughters; all except Mr. Bray
appeared to be injured, and were taken to the hospital.
Some one also threw a child into the street, and he
was caught by one of the persons passing by.

And now up came the firemen with their escape
from Copenhagen Street. Pitching it against the
house, they hurried to the upper windows. From one
of these they brought down a young woman, who was
sadly burnt about the face, and she was sent also
to the hospital. Penetrating still farther amid the
smoke and flame, Arthur Whaley groped about, and
found two lads asleep, and, bearing them out, saved
their lives by means of the escape.

The fire did considerable damage before it was
finally extinguished; but when the stand-pipes were
got fully to work, the flames were quickly subdued.
One of the daughters died from severe burns soon
after her admission to the hospital, and it was
afterwards found that a girl of fifteen had been
unhappily suffocated in bed. But for the bravery of
Whaley, the two little boys might have suffered the
same sad fate.

These true stories of work in the brigade show how
various are the perilous risks to which firemen are
liable. Danger, indeed, meets them at every turn,
and in almost every guise. To cope with these risks
requires instant readiness of resource as well as
knowledge and skill. In times when seconds count

as hours, it is not enough to know what to do, but how to do it with the utmost smartness and efficiency.

Improved appliances will greatly assist the men ; and Commander Wells's horsed escape fully justified expectations soon after its introduction. It can be hurried through the streets at twelve miles an hour, and the wonder is that the brigade used the old hand-driven machine with its slow pace so long. In December, 1898, a horsed escape reached a fire in Goswell Road in a minute from the alarm signalling in St. John's Square fire-station, and saved three lives,—an instance of very smart work that might establish a record, except that great smartness is everywhere the characteristic of the brigade.

Let us, then, look at the story of the fire-escape a little more closely, and also at some of the new improved appliances, such as the new fire-engine floats and the river-service.

CHAPTER XII.

FIRE-ESCAPES AND FIRE-FLOATS.

" Very smart indeed."

The speaker was watching a light van, which had just been whirled into a yard. Light ladders projected horizontally in front of the van, and large wheels hung behind, a few inches above-ground. The machine was glowing in brilliant red paint.

Off jump five men in shining brass helmets.

" Stand by to slip ! " cries one of the men, who is known as No. 1.

Thereupon, another man casts off some fastenings at the head of the van, and controls the ladders until the large wheels touch the roadway ; another man eases away certain tackle ; and yet another, as by a magical touch, brings the ladder to an upright position directly the big red wheels come in contact with the ground, No. 2 man assisting him.

The whole operation is performed with great smartness, and the escape—for the machine is one of Commander Wells's new horsed escapes—is whipped off its van and reared against the house in the proverbial twinkling of an eye.

Such a scene may be witnessed any afternoon at the London Fire-Brigade Headquarters, when the horse-escape drill is being practised ; and the superiority of the new machine over the old seems so obvious, that you exclaim : " I wonder it has not been done before ! "

The men's positions are all assigned to them. The " crew," as it is called, consists of four firemen and a coachman. When hurrying to a fire, No. 1 takes his place on the near side in front, No. 2 is at the brake on the off side, No. 3 at the brake on the near side, while No. 4 takes his seat on the off side.

Arrived at the scene of the fire, each man springs to his appointed duty. When the escape is quite clear, No. 1 goes to the fire, No. 3 is seen busy with the gear, and the coachman is occupied with his horses. He removes them from the van if necessary, and is ready to ride with a message if required to do so.

Moreover, the van carries five hundred feet of hose, and all the necessary gear for using a hydrant at once ; so that water can be thrown on a fire directly, even without the arrival of an engine.

Life-saving is, however, the special use of the escape itself; and looking at it superficially, you will say that the ladder of this machine is not nearly long enough to reach the upper windows of a high house.

But if you watch the men at work, you will see that the ladder can be cleverly and quickly extended to a much greater height.

You will observe that the escape is made on the telescopic principle, and on a sliding carriage; and though when not extended it only measures about 24 feet over all,—as when riding on the van,—yet when the extending gear is set to work, it can be made to reach a height of 50 feet, or more than double its usual length.

This gear for extending the ladder is fitted to the levers on each side, and is easily worked by two men. The 50-feet escapes are in three lengths, the middle ladder being worked by two separate wires, and the top ladder by one wire.

The van carrying the escape is specially built for the purpose; and, as we have seen, the machine can be instantaneously detached, the van being thus free for other uses if necessary.

Not long after his appointment as chief officer in November, 1896, Commander Wells submitted plans which he had designed for new escapes 40 and 50 feet in length, and ladders 70 feet in length. The 40-feet escape was in two lengths, and the others in three lengths; and all of them were designed to be carried on a van of new pattern.

The County Council authorized the chief officer to obtain patents for his invention, and also ordered experimental machines to be made. These proving satisfactory, it was determined to use them; and a considerable number were ordered, the horsed escape being introduced into the brigade in July, 1897. The appliance is lighter than those hitherto in use, and can be manipulated by fewer men with even greater ease.

It has no shoot, or trough, down which a rescued person can be slipped ; and bearing in mind that this operation may prove hazardous, unless the person have sufficient presence of mind to raise and press his arms against either side of the shoot so as to break his fall, there is no reason to regret its absence.

Further, the machine will now be able to reach the scene of action so speedily, and is so amply manned, that the firemen should be able to effect a rescue without the need of a shoot. At the same time, it must be borne in mind that instruction for various patterns of fire-escapes is given at headquarters, and the shoot may be seen in use on some machines there.

The new horsed escape follows a series of life-saving appliances, extending over many years. Ladders of various kinds, of course, form an important feature ; but the necessity of some arrangement whereby the height of the ladders could be rapidly and efficiently extended would, no doubt, stimulate invention ; and various contrivances were devised for this purpose. Further, the need for conveying the machine rapidly to the fire would lead to the ladders being placed on wheels.

Without specifying the various kinds of portable ladders in use, it may be stated that the Metropolitan Brigade came to use one, consisting of a main ladder varying from 32 to 36 feet high, and furnished with a canvas trough along its length. It was doubtless a machine of this sort which was in use when Fireman Ford lost his life at the Gray's Inn Road fire in 1871.

A second ladder, jointed to the first, extended the height 15 feet ; while other ladders in some escapes raised the height to 60 and in some cases to 70 feet. The escape in general use by the brigade in 1889 consisted of a main ladder, having the sides strengthened by patent wire-rope, and finished at the back with a shoot or a trough of uninflammable copper-wire netting. A fly-ladder lay along the main ladder, to which it was jointed, and was raised, when needed, by levers and ropes. A third ladder, known as the "first floor," which could be jointed to the fly-ladder, was placed under the main ladder ; while a fourth could be added, bringing the height up to 60 feet. The fly-ladder could also be instantly detached for separate use if required.

The carriage on which this arrangement of ladders was mounted was comparatively light, and was fitted with springs and high wheels, and two men could move it anywhere.

As we have said, drill for various descriptions of escapes is practised at headquarters ; but the general instructions are that, when running the machine, two men are to be " on the levers," to prevent accident.

There used to be a society to organize the use of fire-escapes. It was called the Royal Society for the Protection of Life from Fire, and was first established in 1836. About seven years later its object was more fully attained, when it was reorganized, and had six escape-stations in the metropolis. In 1866, it possessed no fewer than eighty-five stations, while many lives had been saved, and numerous fires had been attended.

But next year, a municipal fire-brigade having been established, the society handed over its works, and practically made a present of all its plant to the

Metropolitan Board of Works, the Fire-Brigade Act having been passed in 1865. And so once more municipal organization took up and developed what voluntary effort had begun.

Various devices have also been employed to afford escape from the interior of the building. Perhaps the simplest, and yet one of the most effectual, consists of a rope ladder fastened permanently to the window-sill, and rolled up near it ; or a single cord may be used, knotted at points about a foot apart all along its length. Like the rope ladder, the cord may be permanently fastened to the window-sill, and coiled up under the toilet-table, or in any place where it may be out of the way, and yet convenient to hand.

Persons may be lowered by this rope, by fastening them at the end—as, for instance, by tying it under their arms, or placing them in a sack and fastening the rope to it—and then allowing the rope to gradually slip through the hands of the person lowering them. Better still, the rope should be bent round the corner of the window-sill, or round the corner of a bed-post, when the friction on the hands will not be so great, and the gradual descent will be safe-guarded.

In descending alone, a person will find the knots of great assistance in preventing him from slipping down too fast; and he may increase the safety of his descent by placing his feet on the wall as he moves his grip, one hand after the other, on the rope; this arrangement prevents the friction on the hands, which hurried sliding might cause, with its attendant danger of falling.

Permanent fire-escapes are provided in large buildings by means of iron ladders or staircases at the back or sides of the structure, with balconies at each story ; while poles having baskets attached, ropes with weights so that they may be thrown into windows, and various contrivances and combinations of ladders, baskets, nets and ropes, etc., have all been recommended or brought into use during a long course of years. They are designed to afford escape, either from within, or from without, the burning building ; several, however, being for private installation.

Returning, then, to the public improvements in fire extinction, a new and remarkable floating fire-engine was designed about the year 1898, by Messrs. Yarrow & Co. of Poplar, in conjunction with Com-

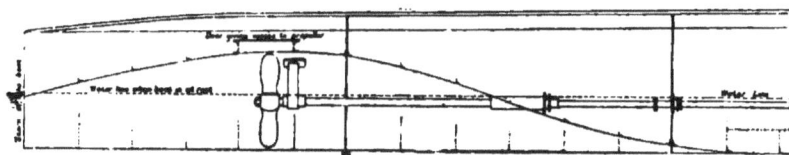

STERN OF YARROW'S FIRE-LAUNCH.

mander Wells, chief of the London Brigade. It was intended for use in very shallow water.

The plan was cleverly based on the lines of the *Heron* type of shallow-draught gunboats constructed for use on tropical rivers. Six of these vessels were built by Messrs. Yarrow for the Admiralty, and two went to the Niger and four to China. The new fire-float design provided for twin-screw propellers fitted in raised pipes, or inverted tunnels, to ensure very light draught combined with high speed, and the consequent power of manœuvring quickly quite near to the shore.

The difficulty of working fire-floats close to the shore in all states of the tide had long troubled the London Brigade, and rendered the best type of vessel for this purpose a matter of much concern. Originally,

9

vessels of comparatively large size were used, containing machinery both for throwing water and for propelling the boat. These vessels, however, were costly to maintain, and could not be effectively used at all states of the tide. Captain Shaw, therefore, separated the fire-engine from the propelling power, using tug-boats which would float in a few feet of water to haul along fire-engine rafts, which could be used quite near to the scene of the fire.

The last of the large vessels disappeared from the brigade in 1890, and the river-service consisted of tugs and floats, the fire-engines or rafts being familiarly called by the latter name. This system, however, did not prove satisfactory ; for, as the chief engineer pointed out, just before the appointment of Commander Wells, tugs being necessary to haul the floats, double the number of river-craft were employed, and there was a consequent increase in cost of maintenance. He suggested that both the propelling and the fire-engine machinery should be united on one vessel, but that it should be of light draught.

The new chief officer was consulted. Now, Commander Wells, who was then thirty-seven years of age, had enjoyed a long experience in the navy ; and, moreover, had been used to torpedo-boats, which of course are comparatively light craft. Entering the Service in 1873, he was second in command of a torpedo-boat destroyer in the Egyptian campaign of 1882, and for three years was second in command of the Torpedo School at Devonport. At the time of his election to the chief officer's post of the London Fire-Brigade, he was senior officer of a torpedo-boat squadron. He had also been second in command of two battleships, and had partly organized the London Naval Exhibition of 1891. He was, therefore, likely to be thoroughly conversant with all the latest types of light-draught navy vessels.

He pointed out the great disparity existing between the brigade's tugs, which required nine feet of water, and the fire-engine floats, which needed only about two feet ; and he prepared a rough plan of a craft on the model of shallow-draught gunboats. The chief engineer approving the plan, a design was prepared by Messrs. Yarrow & Co., in conjunction with Commander Wells.

This design, or one similar to it, is probably destined to revolutionize river fire-engine service. The class of material used would be the same as that employed for building light-draught vessels for her Majesty's Government ; and the method of raising the steam would be, of course, by Yarrow's water tube-boilers, having straight tubes, and raising steam from cold water in fifteen minutes.

The design shows a vessel about 100 feet long by 18 feet beam, and the draught only about 1 foot 7 inches—i.e., five inches less than the previous floats, though containing its own propelling power. The engines, twin-screw and compound, would develop about 180 horse-power, and the speed range from nine to ten knots an hour, while no doubt much higher speed could be obtained if desired.

But the main feature is the ingenious use of the propellers. How can they work in such shallow water ?

Briefly, the propellers operate in the two inverted tunnels, the upper parts of which are considerably above the water-line. When the propellers commence to work, the air is expelled from the tunnels, and is immediately replaced by water. Thus, a large

propeller can be fully immersed, while the vessel itself
is only floating in half or may be a third of the amount
of water in which the propeller is actually working.
The design thus combines maximum speed with
minimum draught.   Sooner or latter, it seems likely
that some such system must be adopted for fire-floats
used in protecting water-side premises ; and so far the
design promises to inaugurate a new era.

The boilers in the design also operate the fire-
engine pumps, which would probably consist of four
powerful duplex " Worthingtons," each throwing five
hundred gallons a minute.   They discharge into a pipe
connected with a large air-vessel, whence a series of
branches issue with valves connected with fire-hose.

But at the top of the large air-vessel stands a water-
tower ladder, the two sides consisting of water-pipes.
At the heads of the pipes are fitted two-inch nozzles,
the direction of which can be varied by moving the
water-ladders from the deck.   Branch-pipes can also
be led underneath the deck to either side of the vessel.
Suitable accommodation is provided for the crew, and
ample deck space is available for working the craft.
She seems likely to give a good account of herself at
any water-side fire to which she might be called.

Concurrently with this new design, arrangements
were made to alter the London river-stations, and to
some extent remodel the river organization.   Previously, there had been five river-stations ; but usually
between fifteen and twenty minutes elapsed after a
fire-alarm was received before a tug got under way
with its raft or float.   This delay was partly owing to
the fact that the men lived at some distance, and also
that a full head of steam was not kept on the tugs.

The chief officer advised that the staff and appliances

of the A and B stations, and also of the C and D
stations, should be amalgamated, and thus a crew
could be always on board and ready to proceed
to a fire at a moment's notice. There would be
four river-stations—*viz*., at Battersea, Blackfriars,
Rotherhithe, and Deptford—from any of which a crew
with appliances could steam at once. The value of
the new arrangement is obvious. Moreover, the staff
of the Blackfriars post are lodged in the large new
fire-engine station at Whitefriars, opened July 21st,
1897, and which is not far from the north of Blackfriars
Bridge.

As, therefore, the nineteenth century closes, we see
the London Brigade, which has formed the model of
so many others in the kingdom, straining every nerve,
not only to maintain its high reputation, but to
develop and to improve its elaborate organization and
its numerous appliances for coping with its terrible
enemy.

But, in the meantime, invention has been busy in
other directions. Fire is so terrible a calamity, and
its risks so great, that ingenuity has been taxed to
the utmost to master it in every way ; and not only
to extinguish it, but to prevent it from occurring at
all. Of a fire, indeed, it may be said that prevention
is better than cure.

What think you of muslin that will not flame, of
ceilings that will pour forth water by themselves, of
glass bottles that break and choke the fire ? What
think you of chemical fire-engines, some so small as
to be easily carried on a man's back? or of curtains
and screens and fabrics that stubbornly refuse to
yield ?

All kinds of contrivances, in short, have been cleverly

designed. Let us now see some in operation. Have
you ever seen a fire choked in a minute ? and how
is it done ?

CHAPTER XIII.

CHEMICAL FIRE-ENGINES. FIRE-PROOFING, OR MUSLIN
THAT WILL NOT FLAME.

WHICH structure will be first extinguished ?

Imagine yourself gazing at two wooden sheds, both
quite filled with combustible materials, and drenched
with petroleum and tar. These are to be fired, and
then one is to be extinguished by water, and the
other by an extinctor, or chemical fire-engine.

" Ready ! "

At the word, the torch is applied, and the first shed
bursts into flames. It soon blazes furiously. A man
steps forward, armed with a hand-pump, such as is
used by the Metropolitan Fire-Brigade, and turns a
jet of water upon it.

Hiss ! squish ! A cloud of steam rises as the water
dashes upon the fire, and still the stream pours on.
Now the fireman pauses to refill his pump with water,
and then again the jet plays on the burning pile.

The fire dims down to a dull red, the flames cease to
shoot upward and outward, and after about five minutes
the conflagration is extinguished. Bravo ! A very
smart piece of work !

But now the second shed is lighted, and blazes fast.
Another man hurries forward. He has a steel cylinder
slung on his back, and in a second, without any pumping, he directs a jet of fluid upon the fire. The flames

die down, the red gives place to blackness, and, in about half the time taken by the other method, the extinctor has completely quenched the fire. How is it done?

Within the steel cylinder is suspended a bottle charged with a powerful acid, probably sulphuric acid—but the secrets of patents must not be revealed. The bottle can be instantaneously broken by a lever or

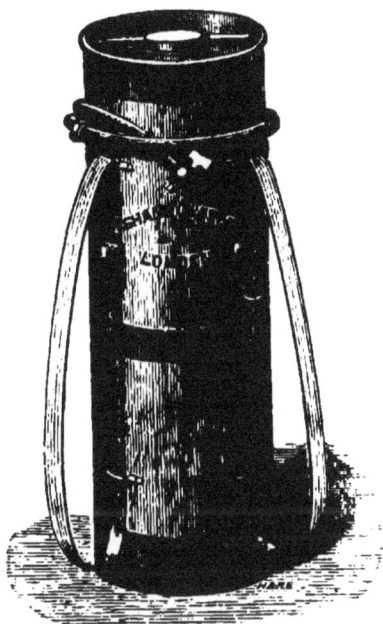

CHEMICAL EXTINCTOR.

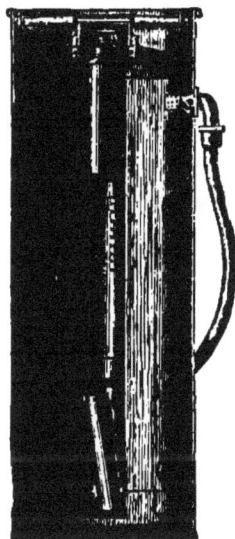

SECTION OF CHEMICAL
FIRE-ENGINE.

weight, and the acid is precipitated into the cylinder, which is filled with an alkaline fluid—perhaps a solution of carbonate of soda. The mixture of these fluids rapidly produces large quantities of carbonic acid gas, which is a great enemy to fire. Moreover, water absorbs the gas easily; and when generated in the cylinder, the expansion of the gas causes a propelling power, varying from seventy to a hundred pounds per

square inch. Consequently, a jet of water propelled
by the gas shoots out a distance varying from thirty
to fifty feet ; and when it reaches the fire, the heat
evaporates the water, and liberates the gas held in
solution, which chokes the fire.

This is the general principle of most chemical fire-
engines. There are several varieties ; but they are, no
doubt, chiefly based on the rapid evolution of carbonic
acid gas. If you find the principle difficult to understand, imagine a soda-water bottle bursting, or the
contents spurting forth if the cork be suddenly removed,
and you will not be so surprised at the stream jetting
forth from an extinctor. Soda-water is, of course,
aërated by being charged with carbonic acid gas.

These chemical extinctors are of all sizes ; they
range from small bottles upward, to large double-tank
machines, and drawn by horses. The small bottles
contain the necessary materials, so arranged that,
when the bottle is thrown down, the gas is generated
and the fire choked. Both Germany and the United
States make large use of chemical fire-engines, some
of which are capable of giving a pressure of a hundred
and forty pounds, and perhaps more, to the square inch.

Cases filled with sulphur, saltpetre, and other
chemicals are sometimes used, which, being ignited,
send forth a choking vapour, stifling all fire in a
confined space ; again, other contrivances discharge
ammoniacal gases and hydrochloric acid.

Extinctors, or fire-annihilators, have been invented
or introduced by several persons. Mr. T. Phillips was
responsible for one in 1849, which generated steam
and carbonic acid. Two or three persons seem to
have had a hand in an apparatus developed by Mr.
W. B. Dick about twenty years later, and patented

April, 1869. This consisted of an iron cylinder furnished with tartaric acid, bicarbonate of soda and water, and generating the carbonic acid gas. The first inventor of this appliance was a Dr. Carlier, who suggested it, or something like it, a few years previously.

About the same time, Mr. James Sinclair introduced his chemical appliance, the firm now being the Harden Star, Lewis, & Sinclair Company. British fire-brigades would not touch the extinctors; but the Americans seized upon them rapidly, and manufactured them largely. At the present time, it is said that there is scarcely a fire-brigade in the States that does not use a chemical fire-engine in some shape or form.

In Britain, the extinctor, either as the hand-grenade bottle or portable cylinder, which latter contains about eight gallons, is largely used by private persons, and is kept in many large establishments. Several provincial fire-brigades have also adopted the appliances in some form or other; but, as a rule, the chemical fire-engine has not been used by the public fire-brigades of the country. Perhaps one reason is, that it is regarded as more suitable for private use, and not as superior to the powerful steam-engines, hydrants, etc., operated so efficiently by trained firemen.

It will be seen that the claims for chemical fire-engines are twofold in character: first, that they themselves supply propelling power for the fluid without pumps—a great consideration for private persons; and, secondly, that the liquid thrown has far greater fire-quenching powers than water.

To the first of these claims, it is possible that fire-brigades, with their numerous hydrants and powerful steam-engines, would pay but little regard; while as to the second claim, only accomplished chemists and impartially-minded persons of wide and varied experience can form a fully-reliable opinion.

At the time of the great Cripplegate fire in London, November, 1897, Americans were very keen in their criticism, much of which was unjust and inaccurate ; but one of their points was the absence of chemical appliances in the London Brigade.

It is, however, fairly open to argument whether the use of such apparatus would have mended matters. Even Americans have by no means abolished the steam fire-engine ; and they have sometimes found that the fire has obtained so firm a hold, that the best they could do was to prevent the flames from spreading. When quantities of inflammable substances are crowded in high and comparatively frail buildings in narrow thoroughfares, you have all the elements of serious fires ; and when once fairly started, it remains to be proved whether a gas-propelled and gas-laden stream would be more efficient than powerful and copious jets of water.

The difficulty would appear to be rather that of directly and quickly reaching the seat of the fire, than of the more or less fire-quenching properties of rival fluids. From the evidence of Mr. John F. Dane at the Cripplegate Fire Enquiry, we may gain some idea why the brigade dislike the chemical fire-engine. He had been twenty-eight years in the brigade—though he had then left the service, and was a consulting fire engineer—but at one fire, where he had found a dense smoke, an hour was occupied in tracing the fire to its source, it being worked upon by hand-buckets. Had he used a chemical fire-engine, it would, no doubt, have been played into the dense smoke, and damaged a thousand pounds' worth of goods, while, after having exhausted the charge, they would not have found the fire subdued. Chemical fire-engines could not be trusted to discharge where wanted.

Many modern structures at the Cripplegate fire were comparatively frail. Iron girders and stone were, no doubt, largely used, and you would naturally think that iron would be fireproof ; but, as a matter of fact, iron may be worse than wood. That is, cast-iron is very liable to split, if suddenly heated or cooled ; and a jet of water playing on a hot cast-iron girder would most likely cause it to collapse at once, and bring down everything it supported in a terrible ruin.

The truth is, therefore, that light iron and stone structures are not nearly so fireproof as they might appear. The difficulty of building fireproof structures has not yet been fully solved, though many suggestions to that end have been made. Wood soaked in a strong solution of tungstate or silicate of soda is rendered uninflammable and nearly incombustible. Silicate of soda is, perhaps, the best. It fuses in the heat, and forms a glaze over the wood, preventing the oxygen in the air from reaching it. But intense heat will overcome it. Whichcord's plan of fire-proofing encases metal girders in blocks of fire-clay ; other systems make great use of concrete. Walls, of course, should be built of brick or stone ; while double iron doors are of great value, as in the case of the warehouses burning at the docks on January 1st, 1866.

At the enquiry into the Cripplegate fire of 1897, Mr. Hatchett Smith, F.R.I.B.A., declared that the well-holes or lighting-areas in the warehouses involved, were a source of danger as constructed, and he recommended that such lighting-areas should be confined by party walls, and sealed with rolled plate-glass or pavement-lights. Windows facing the street should be glazed with double sashes, and external walls should be built with a hollow space of about two inches between them and their plastering, with an automatic water-sprinkler at the top of the hollow space. Such a plan of construction would, he contended, confine the fire to the apartment in which it originated, though it would not extinguish the fire in that room. The flooring Mr. Smith seemed to take for granted would be of concrete and fireproof.

Among other fire precautions, the introduction of the electric light in place of gas may operate as a valuable precautionary measure, especially in theatres and public places ; while a strong iron curtain, to be quickly dropped down between the stage and the auditorium, is also a most valuable precaution.

But all such measures may be largely neutralized by the inflammable contents of the buildings. Some manufactures are remarkably dangerous in this respect, and the extensive storage of certain goods renders even spontaneous combustion probable. Thus, if a well-built fireproof structure contain large quantities of combustible materials, and these burn furiously, the heat evolved may be so great as to conquer almost everything in the building. Indeed, the heat in huge fires is sufficient to melt iron.

Nevertheless, the liability to fire and its destructiveness is much decreased by wise precautionary measures in building, the idea underlying them being that walls, floorings, doors, or what not should be so made as to localize the fire to the apartment in which it originated.

As with buildings, so with clothing. Here is a

piece of muslin. Light it: it will not flame ; it slowly smoulders. But even as the problem of building completely fireproof structures has not been solved, so also the question of fireproof fabrics has not been completely answered.

Progress, however, has been made in that direction. Methods have been adopted whereby the flaming of fabrics can be prevented, and their burning reduced to smouldering.

A solution of tungstate of soda is, perhaps, one of the best chemicals to use for this purpose, for it is believed not to injure the fibre ; but for articles of clothing, borax is better suited, as it does not injure the appearance of the clothes, and it is very effectual in its operation, though it weakens the fibre. Alum, common salt, and sulphate of soda will also diminish or entirely prevent flaming ;. but they tend to weaken the fibre.

A simple experiment illustrates the principle. Any boy who has made fireworks, or dabbled in chemistry, knows that paper—one of the most inflammable of substances—after being soaked in a solution of saltpetre, will not flame, but smoulders quickly at the touch of fire ; hence the name touch-paper, which is used to ignite fireworks.

Some of these salts, then, prevent the fabric from flaming, and also reduce the burning to slow smouldering, the explanation being apparently this,—when the fabric is dipped in solutions of certain salts, tiny crystals are deposited among the fibres on drying, and the inflammability is diminished ; but the effect of the salt upon the fabric has to be considered, and some, such as sulphate of ammonia, will decompose when the goods are ironed with a hot iron.

This necessary operation of the laundry, however, does not affect tungstate of soda ; and all the dresses of a household could be rendered non-inflammable and largely incombustible by dipping them in a solution of this salt. The proportions would be about one pound of the tungstate to a couple of gallons of water. For starched goods, the best way to use the tungstate would be to add one part of it to three parts of the starch, and use the compound in the ordinary manner.

Various methods have been adopted for fire-proofing wood, the strong solution of silicate of soda being one of the best. Asbestos paint is also useful, if it does not peel off, a little trick to which it seems addicted. By another method, the wood is soaked for three hours in a mixture of alum, sulphate of zinc, potash, and manganic oxide, with water and a small quantity of sulphuric acid. But while the inflammability of wood may be removed, it is questionable if it can be rendered entirely incombustible. In short, the problem of absolutely preventing fires by rendering substances perfectly fireproof has yet to be solved, if, indeed, it is capable of solution.

But if fire cannot be entirely prevented, could not some method be devised of automatically quenching the flames directly they break forth ?

Such a method would appear like the prerogative of the good genii of a fairy fable, and beyond the reach of ordinary mortals. But science and human ingenuity which tell so many true " fairy tales " have made some approach to this also. The device is known popularly as " sprinklers," and is contrived somewhat in this way :—

Lines of water-pipes are conducted along the ceilings of the building, and are connected with the water supply through a large tank on the roof. To these pipes, the sprinklers are attached at distances of about ten feet. They are, in some cases, jointed with a soft metal, which melts at a temperature of about 160 degrees ; the valve then falls, and the water is sprayed forth into the apartment.

Other sprinklers are said to act by a thread, which, it is claimed, will burn when the heat reaches a certain temperature and release the water. The essential idea, therefore, is that the heat of the fire shall automatically set free the water to quench it. Such great importance is attached to the use of sprinklers by some insurance-offices, that they offer a large reduction of premiums to those employing them. Again, other sprinklers are not automatic, but require to be set in operation by hand.

Nevertheless, in spite of all these varied precautions, it is unfortunately a platitude to say that fires do occur ; but the point to be noted is, that but for these efforts, they would probably be greater in number and more destructive in their results.

Even when the flames are raging in fury, much may be done by courageous and well-trained men to preserve goods from injury ; and, indeed, much is done by a body of men whose work is perhaps too little known. They pluck goods, as it were, out of the very jaws of the fire, and often while the flames are burning above them. Would you like to know them, and see them at work ?

Behold, then, the black helmets and the scarlet cars of the London Salvage Corps.

CHAPTER XIV.

THE WORK OF THE LONDON SALVAGE CORPS.   THE GREAT CRIPPLEGATE FIRE.

" Where is the fire ? "

" City, sir ; warehouses well alight."

" Off, and away ! "

The horses are harnessed to the scarlet car as quickly as though it were a fire-engine ; the crew of ten men seize their helmets and axes from the wall beside the car, and mount to their places with their officers ; the coachman shakes the reins ; and away dashes the salvage-corps trap to the scene of action.

The wheels are broad and strong ; they do not skid or stick at trifles ; the massy steel chains of the harness shine and glitter with burnishing, and might do credit to the Horse Artillery ; the stout leather helmets and sturdy little hand-axes of the men look as fit for service as hand and mind can make them. Everything was in its right place ; everything was ready for action ; and at the word of command the men were on the spot, and fully equipped in a twinkling.

The call came from the fire-brigade. The brigade pass on all their calls to the salvage corps, and the chiefs of the corps have to use their discretion as to the force they shall send. The public do not as a rule summon the salvage corps. The public summon the fire-brigade, and away rush men and appliances to extinguish the flames and to save life. The primary duty of the salvage corps is to save goods. There is telephonic connection between the brigade and the corps, and the two bodies work together with the utmost cordiality.

We will suppose the present call has come from a big City fire. The chief has to decide at once upon his mode of action. No two fires are exactly alike, and saving goods from the flames is something like warfare with savages—you never know what is likely to happen ; so he has to take in the circumstances of the case at a glance, and shape his course accordingly. Should the occasion require a stronger force, he sends back a message by the coachman of the car ; and in his evidence concerning the great Cripplegate fire, Major Charles J. Fox, the chief officer of the salvage corps, stated that he had seventy men at work at that memorable conflagration.

But see, here is the fire ! Streams of water are being poured on to the flames, and the policemen have hard work to keep back the excited crowd. They give way for the scarlet car, and the salvage men have arrived at the scene of action. Entrance may have to be forced to parts of the burning building, and doors and windows broken open for this purpose.

Crash! crash! The axes are at work. And a minute more the men step within amid the smoke. The firemen may be at work on another floor, and the water to quench the fire may be pouring downstairs in a stream. The noises are often extraordinary. There is not only the rush and roar of the flames, the splashing and gurgling of the water, but the falling of goods, furniture, and may be even parts of the structure itself.

Walls, girders, ceilings may fall, ruins clatter about your ears, clouds of smoke suffocate you, tongues of flame scorch your face ; but if you are a salvage man, in and out of the building you go, while with your brave brethren of the corps you spread out the strong rubber tarpaulins you have brought with you in your trap,

10

and cover up such goods as you find, to preserve them
from damage.   Under these stout coverlets, heaps of
commodities may lie quite safe from injury from water
and smoke.

Overhead you still hear terrible noises.   Safes and
tanks tumble and clatter with dreadful din ; part of
the structure itself, or some heavy piece of furniture,
falls to the ground; dense volumes of water poured
into the windows rain through on to your devoted
head.   But you stick to your post, preserving such
goods as you can in the manner that the chief may
direct.   May be you have to assist in conveying goods
out of reach of the hungry fire, and your training has
taught you how to handle efficiently certain classes
of goods.   Sometimes quantities of water collect in
the basement, doing much damage ; and down there,
splash, splash, you go, to open drains, or find some
means of setting the water free.

On occasion, the men of the salvage corps find
themselves in desperate straits.   At the Cripplegate
fire, one of the corps discovered the staircase in flames,
and his retreat quite cut off.   With praiseworthy
promptitude, he knotted some ladies' mantles together
into a rope, and by this means escaped from a second-
story window to the road below.

On another occasion, Major Fox himself, the chief
of the corps, was rather badly hurt on the hip, when
making his way about a burning building at a fire in
the Borough.   The probability of accident is only too
great, and it was no child's play in training or in
practice which enabled the corps to attain such pro-
ficiency as to carry off a handsome silver challenge
cup at an International Fire Tournament at the
Agricultural Hall in the summer of 1895.

The duties of the salvage corps do not end even when the fire is extinguished. They remain in possession of the premises until the fire-insurance claims are satisfactorily arranged. They do not, however, know which office is paying the particular claims, and all offices unite in supporting the corps. It is, in fact, their own institution, though established under Act of Parliament; and it is not, therefore, like the London Fire-Brigade, a municipal service.

When the brigade was handed over to the Metropolitan Board of Works by the Act of 1865, provision was made for the establishment of a salvage corps, to be supported by the Fire-Insurance Companies, and to co-operate with the brigade. The corps has now five stations, the headquarters—where the chief officer, Major Fox, resides—being at Watling Street in the City. The eastern station is at Commercial Road, Whitechapel; the southern, at Southwark; the northern, at Islington; and the western, at Shaftesbury Avenue.

The force consists of about a hundred men. Their uniform somewhat resembles that of the fire-brigade, being of serviceable dark blue cloth, but with helmets of black leather instead of brass. They are nearly all ex-navy men, excepting the coachmen, some of whom have seen service in the army; indeed, candidates now come from the royal navy direct, but receive a special training for their duties, such as in the handling of certain classes of goods. Their ranks are divided into first, second, and third-class men, with coachmen, and foremen, five superintendents, and one chief officer.

Their work lies largely outside the public eye. They labour, so to speak, under the fire; and it is difficult to estimate the immense quantity of goods they save from damage during the course of the year. Thousands of pounds' worth were saved at the great Cripplegate fire alone in November, 1897. That huge conflagration, which was one of the largest in London since the Great Fire of 1666, may well serve to illustrate the work of the corps.

The alarm was raised shortly before one o'clock mid-day on November 19th, and an engine from Whitecross Street was speedily on the spot. As usual, the salvage corps received their call from the brigade ; and in his evidence at the subsequent enquiry at the Guildhall, Major Fox stated he received the call at headquarters from the Watling Street fire-station, a warehouse being alight in Hamsell Street.

He turned out the trap, and with the superintendent and ten men hurried to the fire. He also ordered other traps to be sent on from the other four stations of the corps, and left the station at two minutes past one.

The Watling Street fire-engine had preceded him ; and when he turned the corner of Jewin Street out of Aldersgate Street, he saw " a bright cone of fire with a sort of tufted top." It was very bright, and he was struck by the absence of smoke. He thought the roof of one of the warehouses had gone, and the flames had got through.

Perceiving the fire was likely to be a big affair, he at once started a coachman back to Watling Street with the expressive instructions to " send everything."

The coachman returned at thirteen minutes past one, so the chief and his party must have arrived at the fire about five minutes past one ; that is, they reached the scene of action in three minutes. The major and superintendent walked down Hamsell Street, and found

upper floors " well alight," and the fire burning downward as well.   It was, in fact, very fierce ; so fierce,
indeed, that he remarked to his companion what a
late call they had received.   The firemen were getting
to work, and he himself proceeded with his salvage
operations.

Believing that some of the buildings were irrevocably
doomed, he did not send his men into these, for the
sufficient reason that he could not see how he could
get the men out again ; but they got to work in other
buildings in Hamsell Street and Well Street, though
the fire was spreading very rapidly.   Many windows
were open, which was a material source of danger,
causing, of course, a draught for the fire.   They shut
some of the windows, and removed piles of goods from
the glass, so that the buildings might resist the flames
as long as possible.   Eventually, the staff of men, now
increased to seventy in number, cleared out a large
quantity of goods, and stacked them on a piece of
vacant ground near Australian Avenue.

In spite of the heat and smoke and flame, in spite
of falling tanks and safes and walls, the men worked
splendidly, and were able to save an immense quantity
of property.

Meantime, the firemen had been working hard.   On
arrival, they found the fire spreading with remarkable
rapidity, and the telephone summoned more and more
assistance.   Commander Wells was at St. Bartholo-
mew's Hospital examining the fire appliances when
he was informed of the outbreak.   He left at once,
and reached Jewin Street about a quarter past one.
Superintendent Dowell was with him ; and on entering
the street, they could see from the smoke that the fire
was large, and that both Hamsell Street and Well

Street were impassable, as flames even then were leaping across both the streets.

Steamers, escapes, and manuals hurried up from all quarters, until about fifty steamers were playing on the flames. Early in the afternoon, the girls employed in a mantle warehouse hastened to the roof in great excitement, and escaped by an adjoining building.

A staff of men soon arrived from the Gas Company's offices ; but the falls of ruins were already so numerous and so dangerous, that they were not able to work effectually.

In fact, the whole of Hamsell Street was before long in flames ; and in spite of all efforts, the fire spread to Redcross Street, Jewin Crescent, Jewin Street, and Well Street. The brigade had arrived with their usual promptitude ; but before their appliances could bring any considerable power to bear, the conflagration was extending fast and fiercely.

The thoroughfares were narrow, the buildings high, and the contents of a very inflammable nature, such as stationery, fancy goods, celluloid articles (celluloid being one of the most inflammable substances known), feathers, silks, etc., while a strong breeze wafted burning fragments hither and thither. Windows soon cracked and broke, the fire itself thus creating or increasing the draught; the iron girders yielded to the intense heat, the interiors collapsed, and the flames raged triumphantly.

In Jewin Crescent, the firemen worked nearly knee-deep in water, and again and again ruined portions of masonry crashed into the roadway. Through the afternoon, engines continued to hurry up, until at five o'clock the maximum number of about fifty was reached. The end of Jewin Street resembled an

immense furnace, while the bare walls of the premises already burnt out stood gaunt and empty behind, and portions of their masonry continued to fall.

Firemen were posted on surrounding roofs and on fire-escape ladders, pouring immense quantities of water on the fire, while others were working hard to prevent the flames from spreading. All around, thousands of spectators were massed, pressing as near as they could. They responded readily, however, to the efforts of the police, and order was well maintained.

This was the critical period of the fire. It still seemed spreading ; in fact, it appeared as though there were half a dozen outbreaks at once. But after six, the efforts of the firemen were successful in preventing it from spreading farther. As darkness fell, huge flames seemed to spurt upward from the earth, presenting a strikingly weird appearance ; they were caused by the burning gas which the workmen had not been able to cut off. Crash succeeded crash every few minutes, as tons of masonry fell ; while in Well Street, at one period a huge warehouse, towering high, seemed wrapped in immense flame from basement to roof.

An accident occurred by Bradford Avenue. Some firemen, throwing water on the raging fire, were suddenly surprised by a terrible outburst from beneath them, and it was seen that the floors below were in flames. To the excited spectators it seemed for a moment as though the men must perish ; but a fire-escape was pitched for them, and amid tremendous cheering the scorched and half-suffocated men slid down it in safety.

Cripplegate Church, too, suffered a narrow escape,

even as it did in the Great Fire of 1666. On both occasions, sparks set fire to the roof, the oak rafters on this occasion being ignited. But the special efforts made by the firemen to save it were happily crowned by success, though it sustained some damage. Also Mr. Nein, one of the churchwardens, assisted by Mr. Morvell and Mr. Capper, posted on the roof, worked hard with buckets to quench the flames.

It was late at night before the official " stop " message was circulated, and eight o'clock next morning before the last engine left. It was found that the area affected by the fire covered four and a half acres, two and a half being burnt out ; and no fewer than a hundred and six premises were involved. Fifty-six buildings were absolutely destroyed, and fifty others burnt out or damaged. Seventeen streets were affected; but happily no lives were lost, though several firemen were burnt somewhat severely. The total loss was estimated at two millions sterling, the insurance loss being put at about half that amount. The verdict, on the termination of the enquiry at the Guildhall on January 12th, 1898, attributed the conflagration to the wilful ignition of goods by some one unknown.

The quantity of water used at this fire was enormous. Mr. Ernest Collins, engineer to the New River Water Company, in whose district the conflagration took place, said that, up to the time when the " stop " message was received, the total reached to about five million gallons. No wonder that the firemen were working knee-deep in Jewin Street. The five million gallons would, he testified, give a depth of about five feet over the whole area. But, further, a large quantity was used for a week or so afterwards, until the conflagration was completely subdued. In addition to the engines, it must be remembered that there were fifty hydrants in the neighbourhood.

These hydrants can, of course, be brought into use without the turncock ; but, as a matter of fact, that official arrived at two minutes past one, the same time as the first engine ; while the fire was dated in the company's return as only breaking out at four minutes to one, and the brigade report their call at two minutes to one.

The water used came from the company's reservoir in Claremont Square, Islington. But this receptacle only holds three and a half million gallons when full. It is, however, connected with another reservoir at Highgate having a capacity of fifteen million gallons, and with yet another at Crouch Hill having when full twelve million gallons. As a matter of fact, these two reservoirs held twenty-five million gallons between them on the day of the fire, and both were brought into requisition, as well as the Islington reservoir. The drain was, however, enormous.

In the course of the first hour, the water in the Islington reservoir actually fell four feet. It never fell lower, however ; for instructions were telegraphed to the authorities at other reservoirs to send on more water, and the supply was satisfactorily maintained,— a striking contrast, indeed, to the Great Fire of 1666, when the New River water-pipes were dry !

It was about nine o'clock when the chief officer of the salvage corps felt able to leave. During the eight hours he had been on duty, his men had saved goods to the value of many thousands of pounds. He had known to some extent the class of goods he would meet with, for the inspectors of the corps make reports from time to time as to the commodities stored in various City warehouses, and he is therefore to some extent prepared. On the following day, the 20th, the corps were occupied in pulling down the tottering walls of the burned-out warehouses which were in a dangerous condition.

This great Cripplegate fire aroused a good deal of attention in the American papers, and certain discussion also arose in England as to water-towers and chemical fire-engines. America is very proud of its well-furnished firemen, and not without cause. Several cities in the States are, indeed, famous for their well-organized and well-equipped fire departments. Let us, then, cross the Atlantic, and see something of the men and their methods in active operation.

We shall find much to interest and to inform us.

CHAPTER XV.

ACROSS THE WATER.

" How can the firemen climb up there ? "

The question may well be asked ; for the tall New York houses seem to reach to the sky.

" Ordinary ladders won't do."

" I guess not," replies the New Yorker. " Why, as far back as 1885, fourteen out of every hundred buildings were too high to be scaled that way. We build tall here."

" Then, how about the fire-escape ? " asks the Englishman.

" Wa'll, iron ladders or steps are permanently fixed to some of the top windows. But the firemen bring their hook-and-ladder ; that is a most valuable contrivance."

Pursuing his enquiries, the Englishman would find that a hook-and-ladder consisted, briefly, of a strong pole, with steps projecting on either side, and a long and stout hook at the top. The fireman can crash this hook through a window, and hang the pole firmly over the window-sill ; the hook, of course, plunging right through into the room.

Climbing up this pole, with another length in his hand, the fireman can hang the second length into the window next above, and so on, up to the very top of the building. He has also a hook in his belt, which he can fasten to the ladder, when necessary, to steady and secure himself. In fact, a well-trained and courageous fireman can climb up the tallest structures by these appliances.

These hooked poles are made of various lengths, ranging from about 10 to 20 feet and more. Some single ladders and extensions reach to over 80 feet ; but it will be seen at once that a succession of, say, ten- or twelve-feet hooked-pole ladders can be easily handled to reach from floor to floor, and that, used by an active and well-trained fireman, it can become a most important appliance for saving life.

St. Louis appears to have been the pioneer city in the use of this apparatus ; but New York and other corporations have followed suit. Since 1883 every candidate for the New York Fire Department must undergo a course of instruction in the use of this and other appliances, and the thorough learning in this work renders them better men for their ordinary duties.

The ladders are wheeled to the fire on a truck

50 feet long, and called a " hook-and-ladder truck." It
carries ladders of different lengths, and also conveys
pickaxes, shovels, battering-rams, fire-extinguishers,
life-lines, etc., and tools for pushing open heavy doors.
The majority of the ladders are placed on rollers, and

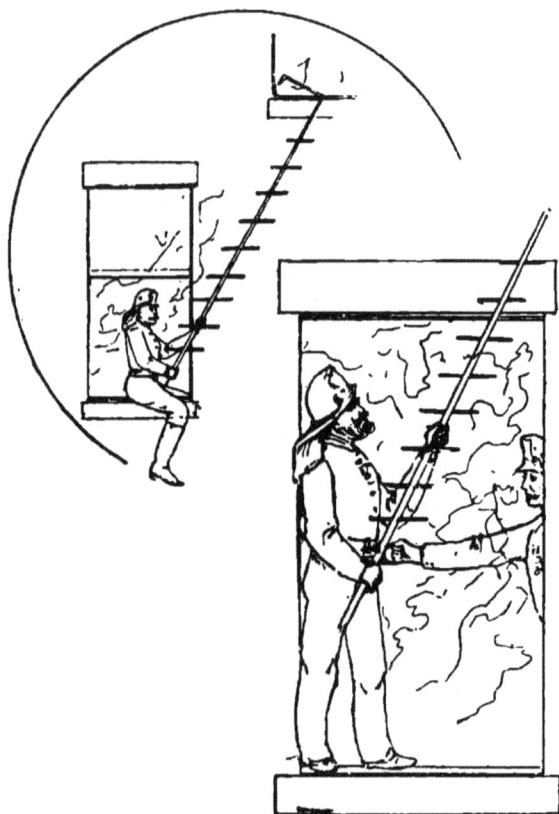

AMERICAN FIRE-LADDERS.

can be removed at once without disturbing those
resting above them.

To some extent, therefore, we might say that the
hook-and-ladder truck with its various appliances
answers to the horsed escape of the London Brigade ;
but, while London firemen make use of the escape as a

point of vantage whence they can discharge water on
the fire, the Americans largely adopt the water-tower.
Indeed, they appear to regard this
apparatus as indispensable for high
business buildings. Briefly, it consists of lengths of pipe, which can
be quickly jointed together, the
lengths being carried on a van, and
varying from about 30 to 50 feet.
When jointed, they can rapidly be
raised to an upright position, the
topmost length having a flexible
pipe and nozzle for the discharge
of the jet of water. This pipe can
be turned in any direction by means
of a wire rope descending below, and
the tower can be revolved by simple
apparatus-gearing. The whole appliance is so arranged that it can be con-
trolled by one man when in action.
The water is supplied by a hose
fastened to the bottom of the tower.

As in England, hydrants are
largely used in the States, and
the steam fire-engine is also, of
course, a very important appliance.
The average American steam fire-
engine generally weighs about three
tons, with water in boiler and men
in their seats on the machine.
The water in the boiler is kept
at steaming-point by a pipe full of
steam passing through it, or boil-
ing water is supplied from a stationary boiler,

AMERICAN FIRE-
LADDERS.

so that on arriving at a fire a working-pressure is obtained. The steam-heating pipe, however, is capable of being instantly disconnected at the sound of the fire-alarm.

The alarm, moreover, is so arranged that the first beat of the gong draws a bolt fastening the horse's halter to the stall. The animals rush to their posts, the firemen slide down poles from the upper stories to the lower, through holes in the floors made for the purpose, and, every one smartly doing his duty, the horses are harnessed, and the engine or apparatus-van is fully ready to start through the open doors before the gong has finished striking—unless it be a very brief alarm.

Four snaps harness the horses.

The animals stand on the ground-floor by the side-walls, facing the wheels of the engines and trucks. The harness is hung over the pole-shaft exactly above the place where the horses will stand, the traces being fastened to the truck ; the hinged collar is snapped round the animal's neck, the shaft-chain is fastened with a snap, and two snaps fix the reins. One shake of the reins by the coachman detaches the harness from the suspenders, and away fly the horses.

Arriving at the fire, the engine is attached to the nearest hydrant, and the delivery-hose is led off to the burning building. The hydrant is probably of the upright kind, standing up above the roadway level, though some cities use the hydrant below-ground, and covered with an iron plate.

But, the water obtained and the engine ready, the method of attacking the fire at close quarters and inside the edifice is adopted if practicable ; and, to accomplish this purpose, the firemen have to fight through blinding and suffocating smoke. For hours they may struggle, well-nigh choked and scorched, though scarce a flash of flame may be visible. To reach the seat of fire, doors are broken down, and even iron shutters opened ; while hose is led upstairs, or down into cellars, in order to quench the flames at their source.

Sometimes, however, on arrival at a fire, the chiefs realize that the conflagration has gained such hold that the firemen's efforts will be most usefully directed to prevent it from spreading. When the water has done its work, the fireman can usually turn it off by a relief-valve without recourse to the engineer, the complete control thus gained tending to prevent unnecessary damage by water.

The American fire-brigades—or departments, as they are called—may be broadly divided into two classes : those of great cities, consisting of a paid staff of officers and men, devoting all their time to the service; and, secondly, those of smaller places, consisting of a staff of unpaid volunteers, pursuing their usual daily avocations, but agreeing to respond to fire-alarms ;—these men, though unpaid, are generally exempt from service as jurymen and militiamen, and sometimes are permitted a slight abatement of taxation. Some brigades, again, consist partly of fully-paid firemen, and partly of volunteers.

Many of these organizations are not only charged with the extinguishment of fires, but also with the regulation of the storage and sale of combustibles, and in some cities with the supervision of building construction. It is claimed that this arrangement has led to a much more economical and efficient administration of this department ; and undoubtedly the fire-

11

brigade has a very lively interest in the security and
stability of buildings. The firemen's efforts to improve
them rank as valuable precautions and preventives of
fire.

It is also claimed that some of the American fire
departments, as, for instance, that of New York, are
among the best in the world, and their engines superior
in size and capacity and greater in number than those
of other lands. On the other hand, the laws regulating
the prevention of fires are said to have been less
stringent than those obtaining in some other countries.

The terrible fire in New York in 1835, when the
loss reached three million pounds, led to the development of the fire-service and of apparatus ; and prizes
were offered for designs for steam fire-engines.
Cincinnati appears to have taken the lead at first ; ·
but the New York Fire Department is now regarded
as one of the most perfect. It is under the control of
three commissioners, whom the mayor appoints ; and
it has substantially a military organization.

The paid brigades are usually divided into companies,
varying from six to twelve individuals, including both
officers and men. A company may be supplied with
a steam fire-engine and tender for hose, or with a
chemical fire-engine, such being called " engine companies " ; or with a hook-and-ladder truck and horses,
called " hook-and-ladder companies " ; or with hose-cart
only and horses, called " hose companies." A hose-
cart will carry nearly a thousand feet of hose, as well
as tools for use at fires and half a dozen firemen ;
while some of them also convey short scaling-ladders.

A water-tower is sometimes placed with an engine
or a hook-and-ladder company ; and, again, these two
companies are occasionally brigaded together. Further,

many cities are arranged into company districts, the captain of each company taking general control over all material, and the enforcement of the laws connected with his department. In some cities, the companies are combined into battalions under a chief of battalion, the highest officer commanding the whole being known as the chief of the department, or may be rejoicing in the imposing title of the Fire-Marshal.

Some of the larger cities have shown their wisdom in appointing their firemen for life, including the highest officers, dismissal only taking place on misconduct ; but in others the baneful practice is followed of dismissing at least the chief officials after a change in local politics—a plan which does not conduce to great efficiency and discipline. New York has abandoned this policy since 1867.

In arranging sites for the fire companies, the principle pursued is to distribute small companies with different appliances over as wide an area as possible, instead of concentrating men and appliances at certain central points. By thus placing companies separately, it is believed that a larger area is served in the same time than by concentrating them together. On the occurrence of large fires, when many companies are called out, distant companies are called from various points, like reserves, to take the places of some of those in action, to meet calls that may arise in the same districts ; while at some stations, or company houses, the men are divided into two sections with duplicate apparatus, so that, while one responds instantly to the first call, the second at once prepares to answer any subsequent alarm.

Among the apparatus used is the jumping-sheet, designed as a last attempt to save life ; circular rope

nets some 15 feet in diameter being carried on the
tenders and trucks in New York ; while their canvas
sheets have rope handles.   Light chemical fire-engines
are also largely used in small places and in the
suburbs of large cities, the lightness of the machine
being, no doubt, a great recommendation.   An efficient
pattern is the double-tank engine, one tank of which
can be replenished while the other is being discharged.
The tanks contain a mixture producing carbonic acid
gas, which is a great foe to fire.   The gas is absorbed
by water, and as it expands causes a great pressure,
sufficient to force the fluid through hose, and throw
it a distance of about a hundred and fifty feet.
When the water reaches the flames, the gas held in
solution is liberated by the heat and chokes the fire.
The mixture will not freeze, even when the temperature
falls to zero ; it is thus always ready.   The machine
is light, and contains its own propulsive force for the
water ; so that we cannot wonder it is widely adopted.

Similar apparatus throws hydrochloric acid and
ammoniacal gases, but opinions differ as to their
utility ; for though efficient fire-quenchers, yet a small
portion only of the gas appears to be carried by the
fluid and actually reaches the flames.

Another piece of apparatus is a hose-hoister.   For
using hose on very high buildings, and also, indeed, for
hoisting ladders to great heights, a simple appliance
has been devised, consisting essentially of a couple of
rollers in a frame ; the rope, of course, runs over the
rollers to hoist the hose, but the frame is shaped to
adjust itself to the coping or cornice of the wall.

For cities on rivers, fire-boats are in use, some being
fitted with twin-screw propellers, and the crew being
sometimes berthed on shore ; while, lastly, as in

England, the fire-alarm telegraph forms a marked feature of the American system. The alarms are fitted with keyless doors, and the telephone is also largely in use. When, however, the keyless door of the alarm is opened by its handle, a gong sounds on the spot, attracting attention, and preventing, it is intended, wrongful interference with the alarm. When the door is opened, the call for the fire company is then sent.

As for the horses, they are regularly trained in New York. They are accepted on trial at the dealer's risk, and placed in a training-stable ; here they grow accustomed to the startling clang of the alarm-gong, to the use of the harness, and to being driven in an engine or ladder-truck.

Passing through these trials satisfactorily, the animal is promoted to service in a company ; and if, after a time, a good report is forthcoming as to activity, intelligence, etc., it is bought in and placed on the regular staff. Then it is given a registered number, which is stamped in lead, and worn round the creature's neck. A record is kept of each horse, the average term of service working out at about six years.

Some horses are so highly trained that they will stand in their stalls unfastened ; others are simply tethered by a halter-strap, a bolt in the stall-side holding a ring in the strap. It is this bolt which is withdrawn by the first beat of the fire-alarm, instantly releasing the horse.

Fire-horses often develop heart disease, as a result of the excitement of their work, and sudden deaths sometimes occur. When beginning to show signs of varying powers or of unfitness for their exciting duties, the horses are sold out of the service, being still useful for many other purposes.

It was of one such that Will Carleton wrote in stirring verse. The old fire-horse was sold to a worthy milkman, and instead of the exciting business of rushing to fires came the useful occupation of taking around milk.

But one day the old horse heard the exciting cry it knew so well. The rush of the fire-horses sounded near ; the engine rattled past. The influence was too strong. Regardless of the milk, the old fire-horse started forward ; his eye gleamed with the old excitement ; no effort could restrain him, and he swept along to the fire, with the lumbering milkcart behind. Over fell the cans ; the milk splashed all over the streets ; but on and on tore the steed, until he actually came in front of the fire-horses, and kept the lead. Then, when he reached the fire, he halted, moped, and presently fell in the street, and died. He was game to the last.

This glance at the American fire departments indicates the great excellence which many of them have reached. The remarkable efficiency is found both in organization and in appliances, and it no doubt invites comparison with British fire-brigades. If so, Britain has nothing to fear. Such comparison, if superficial, is little worth ; and if exhaustive, would consider all the varying circumstances of each country, and would discover great merit on both sides.

Thus, the immense height of the American edifices, no doubt, renders the hook-and-ladder a most valuable appliance ; but buildings in Britain, under the present Acts, are not likely to tower so high ; and the improved fire-escapes so deftly handled by British firemen yield as good, or even better, results for the work they have to do. The question of the chemical fire-engine is for experts and experience to decide ; and whether, with its fumes and its gases, it is really superior under all circumstances, and whether it will ever supersede the water-engine for all purposes, the twentieth century may reveal.

We conclude that absolute superiority cannot be claimed by any one country. The truth is, that the means of fighting fire have been developed to very great excellence in many places ; and when we consider the high courage and efficient training of the men, and the valuable improvements and great usefulness of the various engines and appliances employed, we may truly regard this immense development as one of the wonders of the modern world.

THE END.

Printed by Hazell, Watson, & Viney, Ld., London and Aylesbury.

IN THE SAME SERIES.

THE WORLD'S WONDERS SERIES of *Popular Books treats of the present-day wonders of Science and Art. They are well written, printed on good paper, and fully illustrated. Crown 8vo, 160 pages. Handsome Cloth Cover.* 1s. 6d. *each.*

MARVELS OF ANT LIFE.
By W. F. KIRBY, F.L.S., F.E.S., of the Natural History Museum, South Kensington.

THE ROMANCE OF THE SAVINGS BANKS.
By Arch. G. BOWIE.

THE ROMANCE OF THE POST OFFICE : Its Inception
and Wondrous Development. By Arch. G. BOWIE.

THE ROMANCE OF GLASS-MAKING : A Sketch of the
History of Ornamental Glass. By W. GANDY.

TRIUMPHS OF THE PRINTING PRESS.
By WALTER JERROLD.

ASTRONOMERS AND THEIR OBSERVATIONS
By LUCY TAYLOR. With Preface by W. THYNNE LYNN, B.A F.R.A.S.

MARVELS OF METALS.
By F. M. HOLMES.

MINERS AND THEIR WORKS UNDERGROUND.
By F. M. HOLMES.

CELEBRATED MECHANICS AND THEIR ACHIEVE-
MENTS.
By F. M. HOLMES.

CHEMISTS AND THEIR WONDERS.
By F. M. HOLMES.

ENGINEERS AND THEIR TRIUMPHS.
By F. M. HOLMES.

ELECTRICIANS AND THEIR MARVELS.
By WALTER JERROLD.

MUSICIANS AND THEIR COMPOSITIONS.
By J. R. GRIFFITHS.

NATURALISTS AND THEIR INVESTIGATIONS.
By GEORGE DAY, F.R.M.S.

LONDON : S. W. PARTRIDGE & CO., 8 & 9, PATERNOSTER ROW.

Catalogue

OF

S. W. PARTRIDGE & CO.'S
POPULAR ILLUSTRATED BOOKS.

CLASSIFIED ACCORDING TO PRICES.

NEW BOOKS AND NEW EDITIONS ARE MARKED WITH AN ASTERISK.

5s. each.

*A Hero King: A Romance of the Days of Alfred the Great. By Eliza F. Pollard, Author of "A Gentleman of England," "The White Dove of Amritzir," etc. Large Crown 8vo. Frontispiece. Cloth extra.

The Dacoit's Treasure; or, In the Days of Po Thaw. £200 Prize Story of Burmese Life. By Henry Charles Moore. Illustrated by Harold Piffard. Large Crown 8vo. Cloth extra, gilt top.

A Gentleman of England. A Story of the Time of Sir Philip Sidney. By Eliza F. Pollard, Author of "The White Dove of Amritzir," "Roger the Ranger," etc. Large Crown 8vo. Cloth extra, gilt top.

Pilgrims of the Night. By Sarah Doudney, Author of "A Romance of Lincoln's Inn," "Louie's Married Life," etc. Frontispiece. Large Crown 8vo. Cloth extra, gilt top.

By G. MANVILLE FENN.

Illustrated by W. RAINEY, R.I., F. W. BURTON, etc.

*Jungle and Stream; or, The Adventures of Two Boys in Siam. Large Crown 8vo. Illustrated. Cloth extra.

Cormorant Crag: A Tale of the Smuggling Days. By G. Manville Fenn. Second Edition. Illustrated. Large Crown 8vo. Cloth extra, gilt top.

In Honour's Cause: A Tale of the Days of George the First. By George Manville Fenn, Author of "Cormorant Crag," etc. Large Crown 8vo. Illustrated. Cloth extra, gilt top.

Steve Young; or, The Voyage of the "Hvalross" to the Icy Seas. Large crown 8vo. Fully Illustrated. Cloth extra, gilt top.

3s. 6d. each.

*Grand Chaco (The). By G. Manville Fenn. Large
Crown 8vo. 416 pages. Illustrated. Cloth extra.

*First in the Field: A Story of New South Wales. By
the same Author. Large Crown 8vo. 416 pages. Illustrated. Cloth
extra.

*Through Fire and Storm: Stories of Adventure and
Peril. By G. A. Henty, G. Manville Fenn, and John A. Higginson.
Crown 8vo. 320 pages. Frontispiece. Cloth extra, gilt top.

Skeleton Reef (The). A Sea Story. By Hugh St.
Leger, Author of "An Ocean Outlaw,' etc. Large Crown 8vo.
Frontispiece. Cloth extra, gilt top.

Scuttling of the "Kingfisher" (The). By Alfred E.
Knight, Author of "Victoria: Her Life and Reign." Frontispiece.
Large Crown 8vo. Cloth extra, gilt top.

Missing Million (The): A Tale of Adventure in Search
of a Million Pounds. By E. Harcourt Burrage, Author of "Whither
Bound?" Frontispiece. Large Crown 8vo. Cloth extra, gilt top.

Come, Break Your Fast: Daily Meditations for a
Year. By Rev. Mark Guy Pearse. 544 pages. Large Crown 8vo.
Cloth extra.

Hymn Writers and their Hymns. By Rev. S. W.
Christophers. 390 pages. Crown 8vo. Cloth extra.

Pilgrim's Progress (The). By John Bunyan. Illustrated
with 55 full-page and other Engravings, drawn by Frederick Barnard,
J. D. Linton, W. Small, and engraved by Dalziel Brothers. Crown
4to. Cloth extra, 3s. 6d. (Gilt edges, 5s.)

Romance of Lincoln's Inn (A). By Sarah Doudney,
Author of "Louie's Married Life." Crown 8vo. Illustrated. Cloth.

Story of the Bible (The). Arranged in Simple Style
for Young People. One Hundred Illustrations. Demy 8vo. Cloth
extra, 3s. 6d. (Gilt edges, bevelled boards, 4s. 6d.)

Six Stories by "Pansy." Imperial 8vo. 390 pages.
Fully Illustrated and well bound in cloth, with attractive coloured
design on cover, and Six complete Stories in each Vol. Vols. 1, 2,
3, 4, and 5, 3s. 6d. each.

Two Henriettas (The). By Emma Marshall, Author of
"Eaglehurst Towers," etc. Illustrated. Large Crown 8vo. Cloth
extra, gilt top.

White Dove of Amritzir (The): A Romance of Anglo-
Indian Life. By Eliza F. Pollard, Author of "Roger the Ranger,"
etc. Large Crown 8vo. Illustrated. Cloth extra, gilt top.

2s. 6d. each.
"ROMANCE OF COLONIZATION."

Special attention is requested to this well-written and up-to-date Series of books on the development of British Colonization from its commencement to the present day.

Crown 8vo. Frontispiece. 320 pages. Cloth extra, 2s. 6d. each.

*IV.—**Canada**: Its Rise and Progress. By G. Barnett Smith.

I.—The United States of America to the Time of the Pilgrim Fathers. By G. Barnett Smith.

II.—The United States of America to the Present Day. By G. Barnett Smith.

III.—India. By Alfred E. Knight.

***The Son of Ingar:** A Story of the Days of the Apostle Paul. By K. P. Woods. Frontispiece. Crown 8vo. Cloth extra.

Victoria: Her Life and Reign. By Alfred E. Knight. New Edition. Large Crown 8vo. 320 pages. Cloth extra, 2s. 6d. ; fancy cloth, gilt edges, 3s. 6d. ; half morocco, or half calf, marbled edges, net 7s. 6d. ; full morocco, or calf, gilt edges, net 10s. 6d.

John: A Tale of the Messiah. By K. Pearson Woods. Frontispiece. Crown 8vo. Cloth extra.

Brought to Jesus: A Bible Picture Book for Little Readers. Containing Twelve large New Testament Scenes, printed in colours, with appropriate letterpress by Mrs. G. E. Morton. Size, 13½ by 10 inches. Handsome coloured boards with cloth back.

Bible Pictures and Stories. Old and New Testament. In one Volume. Bound in handsome cloth, with eighty-nine full-page Illustrations by Eminent Artists.

Light for Little Footsteps; or, Bible Stories Illustrated. By the Author of "A Ride to Picture Land," etc. With beautiful coloured Cover and Frontispiece. Full of Pictures.

Potters: Their Arts and Crafts. Historical, Biographical, and Descriptive. By John C. Sparkes (Principal of the Royal College of Art, South Kensington Museum), and Walter Gandy. Crown 8vo. Copiously Illustrated. Cloth extra, 2s. 6d. ; art linen, gilt edges, 3s. 6d.

Story of Jesus. For Little Children. By Mrs. G. E. Morton, Author of "Wee Donald," etc. Many Illustrations. Imperial 16mo.

Sunshine for Showery Days: A Children's Picture-Book. By the Author of "A Ride to Picture Land," etc. Size, 15½ by 11 inches. Coloured Frontispiece, and 114 full-page and other Engravings. Coloured paper boards, with cloth back.

Spiritual Grasp of the Epistles (The); or, an Epistle a-Sunday. By Rev. Charles A. Fox, Author of "Lyrics from the Hills," etc. Small Crown 8vo. Cloth boards. (Not illustrated.)

Upward and Onward. A Thought Book for the Threshold of Active Life. By S. W. Partridge. (Fourteenth Thousand.) Cloth boards, 2s. 6d. (Not Illustrated.)

2s. 6d. each.

THE "RED MOUNTAIN" SERIES.

Crown 8vo. 320 Pages. Illustrated. Handsomely bound in cloth boards. 2s. 6d. each.

***Norcliffe Court.** By John W. Kneeshaw, Author of "A Black Shadow," "From Dusk to Dawn," etc.

***The Inca's Ransom :** A Story of the Conquest of Peru. By Albert Lee, Author of "The Black Disc," "The Prince's Messenger," etc.

***Adventures of Mark Paton (The).** By Charles J. Mansford, Author of "Shafts from an Eastern Quiver," etc.

***Adventures of Don Lavington (The).** By G. Manville Fenn. Illustrated. Large Crown 8vo. Cloth extra.

***Crystal Hunters (The) :** A Boy's Adventures in the Higher Alps. By G. Manville Fenn. Illustrated. Large Crown 8vo. Cloth extra.

***In Battle and Breeze.** Sea Stories by G. A. Henty, G. Manville Fenn, and E. Harcourt Burrage.

A Polar Eden : or, The Goal of the "Dauntless." By Charles R. Kenyon, Author of "The Young Ranchman," etc.

By Sea-Shore, Wood, and Moorland : Peeps at Nature. By Edward Step, Author of "Plant Life," etc.

Eaglehurst Towers. By Emma Marshall, Author of "Fine Gold," etc.

Eagle Cliff (The) : A Tale of the Western Isles. By R. M. Ballantyne, Author of "Fighting the Flames," "The Lifeboat," etc.

Edwin, The Boy Outlaw ; or, The Dawn of Freedom in England. A Story of the Days of Robin Hood. By J. Frederick Hodgetts, Author of "Older England," etc.

England's Navy : Stories of its Ships and its Services. With a Glance at some Navies of the Ancient World. By F. M. Holmes, Author of "Great Works by Great Men," etc.

Green Mountain Boys (The) : A Story of the American War of Independence. By Eliza F. Pollard, Author of "True unto Death," "Roger the Ranger," etc., etc.

Great Works by Great Men : The Story of Famous Engineers and their Triumphs. By F. M. Holmes.

Lady of the Forest (The). By L. T. Meade, Author of "Scamp and I," "Sweet Nancy," etc.

Leaders Into Unknown Lands : Being Chapters of Recent Travel. By A. Montefiore-Brice, F.G.S., F.R.G.S. Maps, etc.

Lion City of Africa (The) : A Story of Adventure. By Willis Boyd Allen, Author of "The Red Mountain of Alaska," etc.

Mark Seaworth : A Tale of the Indian Archipelago. By W. H. G. Kingston, Author of "Manco, the Peruvian Chief."

Manco, The Peruvian Chief. By W. H. G. Kingston. New Edition. Illustrated by Launcelot Speed.

2s. 6d. each.

THE "RED MOUNTAIN" SERIES (*continued*).

Olive Chauncey's Trust. By Mrs. E. R. Pitman, Author of "Lady Missionaries in Foreign Lands."

Roger the Ranger: A Story of Border Life among the Indians. By Eliza F. Pollard, Author of "Not Wanted," etc.

Red Mountain of Alaska (The). By Willis Boyd Allen, Author of "Pine Cones," "The Northern Cross," etc.

Slave Raiders of Zanzibar (The). By E. Harcourt Burrage, Author of "Gerard Mastyn," "Whither Bound?" etc.

Spanish Maiden (The): A Story of Brazil. By Emma E. Hornibrook, Author of "Worth the Winning," etc.

True unto Death: A Story of Russian Life and the Crimean War. By Eliza F. Pollard, Author of "Roger the Ranger."

Vashti Savage: The Story of a Gipsy Girl. By Sarah Tytler.

Whither Bound? A Story of Two Lost Boys. By Owen Landor. With Twenty Illustrations by W. Rainey, R.I.

Young Moose Hunters (The): A Backwoods-Boy's Story. By C. A. Stephens. Profusely Illustrated.

2s. each.

The Friends of Jesus. Illustrated Sketches for the young, of the Twelve Apostles, the Family at Bethany, and other of the earthly friends of the Saviour. Small 4to. Cloth extra.

Animals and their Young. By Harland Coultas. With Twenty-four full-page Illustrations by Harrison Weir. Fcap. 4to Cloth gilt, bevelled boards.

Domestic Pets: Their Habits and Treatment. Anecdotal and Descriptive. Full of Illustrations. Fcap. 4to. Cloth extra.

Our Dumb Companions. By Rev. T. Jackson, M.A. One Hundred and Twenty Illustrations. Fcap. 4to. Cloth extra.

*Bible Picture Roll. Containing a large Engraving of a Scripture Subject, with letterpress for each day in the month.

Sunny Teachings. (New Series.) A Bible Picture Roll containing Twelve beautifully Coloured Scripture Pictures selected from the New Testament. Mounted on roller.

Young Folk's Bible Picture Roll (The). Contains Twelve beautifully Coloured Pictures of Bible Subjects. Printed on good paper, and mounted on roller, with cord for hanging up.

Natural History Picture Roll. Consisting of Thirty-one Illustrated Leaves, with simple large-type Letterpress, suitable to hang up in the Nursery, Schoolroom, etc.

2s, each.

THE HOME LIBRARY.

Crown 8vo. 320 pages. Handsome Cloth Cover. Illustrations.

*Clouds that Pass. By E. Gertrude Hart.

*A Child of Genius. By Lily Watson.

*Out of the Deep. By E. Harcourt Burrage.

*Miss Elizabeth's Niece. By M. S. Haycraft.

*Through the Crucible. By J. Harwood Panting.

*More Precious than Gold. By Jennie Chappell.

*John Halifax, Gentleman. By Mrs. Craik. New Edition. 540 pages.

Ailsa's Reaping; or, Grape-Vines and Thorns. By Jennie Chappell.

Avice : A Story of Imperial Rome. By Eliza F. Pollard.

Brownie; or, The Lady Superior. By Eliza F. Pollard.

Ben-Hur. By L. Wallace.

Better Part (The). By Annie S. Swan.

Bunch of Cherries (A). By J. W. Kirton.

Cousin Mary. By Mrs. Oliphant, Author of "Chronicles of Carlingford," etc.

Dr. Cross; or, Tried and True. By Ruth Sterling.

Dorothy's Training; or, Wild-Flower or Weed? By Jennie Chappell.

Edith Oswald; or, Living for Others. 224 pages. By Jane M. Kippen.

For Honour's Sake. By Jennie Chappell.

Gerard Mastyn; or, The Son of a Genius. By E. Harcourt Burrage.

Gerald Thurlow; or, The New Marshal. By T. M. Browne.

Honor : A Nineteenth Century Heroine. By E. M. Alford.

Household Angel (The). By Madeline Leslie.

Her Saddest Blessing. By Jennie Chappell.

Jacques Hamon ; or, Sir Philip's Private Messenger. By Mary E. Ropes.

Living It Down. By Laura M. Lane.

Louie's Married Life. By Sarah Doudney.

Madeline; or, The Tale of a Haunted House. By Jennie Chappell.

Morning Dew-Drops. By Clara Lucas Balfour.

Mark Desborough's Vow. By Annie S. Swan.

Mick Tracy, the Irish Scripture Reader. By the Author of "Tim Doolan, the Irish Emigrant."

Naomi; or, The Last Days of Jerusalem. By Mrs. Webb.

2s. each.

THE HOME LIBRARY *(continued).*

Pilgrim's Progress (The). By John Bunyan. 416 pages. 47 Illustrations.

Petrel Darcy; or, In Honour Bound. By T. Corrie.

Strait Gate (The). By Annie S. Swan.

Tangled Threads. By Esmá Stuart.

Tom Sharman and his College Chums. By J. O. Keen, D.D.

Uncle Tom's Cabin. By Harriet Beecher Stowe.

Village Story (A). By Mrs. G. E. Morton, Author of " The Story of Jesus," etc.

Without a Thought; or, Dora's Discipline. By Jennie Chappell.

Way in the Wilderness (A). By Maggie Swan.

By " PANSY."

Chrissy's Endeavour.	Ruth Erskine's Crosses.
Three People.	Ester Ried.
Four Girls at Chautauqua.	Ester Ried Yet Speaking.
An Endless Chain.	Julia Ried.
The Chautauqua Girls at Home.	The Man of the House.
Wise and Otherwise.	

Over 385,000 of these volumes have already been sold.

1s. 6d. each.

THE "WORLD'S WONDERS" SERIES.

A Series of Popular Books treating of the present-day wonders of Science and Art. Well written, printed on good paper, and fully illustrated. Crown 8vo, 160 pages. Handsome Cloth Cover.

*Marvels of Ant Life. By W. F. Kirby, F.L.S., F.E.S., of the Natural History Museum, South Kensington.

*The Romance of the Savings Banks. By Arch. G. Bowie.

*The Romance of Glass-Making: A Sketch of the History of Ornamental Glass. By W. Gandy.

The Romance of the Post Office: Its Inception and Wondrous Development. By Arch. G. Bowie.

Marvels of Metals. By F. M. Holmes.

Miners and their Works Underground. By F. M. Holmes.

Triumphs of the Printing Press. By Walter Jerrold.

Astronomers and their Observations. By Lucy Taylor. With Preface by W. Thynne Lynn, B.A., F.R.A.S.

Celebrated Mechanics and their Achievements. By F. M. Holmes.

Chemists and their Wonders. By F. M. Holmes.

Engineers and their Triumphs. By F. M. Holmes.

Electricians and their Marvels. By Walter Jerrold.

Musicians and their Compositions. By J. R. Griffiths.

Naturalists and their Investigations. By George Day, F.R.M.S.

1s. 6d. each.
NEW SERIES OF MISSIONARY BIOGRAPHIES.

Crown 8vo. 160 pages. Cloth extra. Fully Illustrated.

Amid Greenland Snows; or, The Early 'By
History of Arctic Missions.   } Jesse
Bishop Patteson, the Martyr of Melanesia. } Page.

Captain Allen Gardiner: Sailor and Saint. By
Jesse Page, Author of "Japan, its People and Missions," etc.

Congo for Christ (The): The Story of the Congo Mission.
By Rev. J. B. Myers, Author of "William Carey," etc.

David Brainerd, the Apostle to the North
American Indians. By Jesse Page.

Henry Martyn: His Life and Labours—Cam-
bridge, India, Persia. By Jesse Page.

Japan: Its People and Missions. By Jesse Page.

John Williams, the Martyr Missionary of Poly-
nesia. By Rev. James J. Ellis.

James Calvert; or, From Dark to Dawn in Fiji.
By R. Vernon.

Lady Missionaries in Foreign Lands. By Mrs. E.
R. Pitman, Author of "Vestina's Martyrdom," etc.

Madagascar: Its Missionaries and Martyrs. By
William J. Townsend, Author of "Robert Morrison," etc.

Missionary Heroines in Eastern Lands. By Mrs.
E. R. Pitman, Author of "Lady Missionaries in Foreign Lands."

Reginald Heber, Bishop of Calcutta, Author of
"From Greenland's Icy Mountains." By A. Montefiore, F.R.G.S.

Robert Moffat, the Missionary Hero of Kuruman.
By David J. Deane.

Samuel Crowther, the Slave Boy who became
Bishop of the Niger. By Jesse Page.

Thomas Birch Freeman, Missionary Pioneer to
Ashanti, Dahomey, and Egba. By Rev. John Milum, F.R.G.S.

Thomas J. Comber, Missionary Pioneer to the
Congo. By Rev. J. B. Myers, Association Secretary, Baptist
Missionary Society.

Tiyo Soga: The Model Kaffir Missionary. By H. T.
Cousins, Ph.D., F.R.G.S.

William Carey, the Shoemaker who became the
Father and Founder of Modern Missions. By Rev. J. B. Myers.

1s. 6d. each.
NEW POPULAR BIOGRAPHIES.

Crown 8vo. 160 pages. Maps and Illustrations. Cloth extra.

*Four Noble Women and their Work: Sketches of the Life and Work of Frances Willard, Agnes Weston, Sister Dora, and Catherine Booth. By Jennie Chappell.

Canal Boy who became President (The). By Frederic T. Gammon. Twelfth Edition. Thirty-fourth Thousand.

David Livingstone: His Labours and His Legacy. By Arthur Montefiore-Brice, F.G.S., F.R.G.S.

Florence Nightingale, the Wounded Soldier's Friend. By Eliza F. Pollard.

Four Heroes of India: Clive, Warren Hastings, Havelock, Lawrence. By F. M. Holmes.

Fridtjof Nansen: His Life and Explorations. By J. Arthur Bain.

General Gordon, the Christian Soldier and Hero. By G. Barnett Smith.

Gladstone (W. E.): England's Great Commoner. By Walter Jerrold. With Portrait and thirty-eight other Illustrations.

Heroes and Heroines of the Scottish Covenanters. By J. Meldrum Dryerre, LL.B., F.R.G.S.

John Knox and the Scottish Reformation. By G. Barnett Smith.

Michael Faraday, Man of Science. By Walter Jerrold.

Philip Melancthon: The Wittemberg Professor and Theologian of the Reformation. By David J. Deane, Author of "Two Noble Lives," etc.

Sir Richard Tangye ("One and All"). An Autobiography. With Twenty-one Original Illustrations by Frank Hewitt. (192 pages.)

Sir John Franklin and the Romance of the North-West Passage. By G. Barnett Smith.

Slave and His Champions (The): Sketches of Granville Sharp, Thomas Clarkson, William Wilberforce, and Sir T. F. Buxton. By C. D. Michael.

Stanley (Henry M.), the African Explorer. By Arthur Montefiore-Brice, F.G.S., F.R.G.S.

Spurgeon (C. H.): His Life and Ministry. By Jesse Page.

Two Noble Lives: JOHN WICLIFFE, the Morning Star of the Reformation; and MARTIN LUTHER, the Reformer. By David J. Deane. (208 pages.)

Through Prison Bars: The Lives and Labours of John Howard and Elizabeth Fry, the Prisoner's Friends. By William H. Render.

William Tyndale, the Translator of the English Bible. By G. Barnett Smith.

Over 430,000 of these popular volumes have already been sold.

1s. 6d. each.

<u>THE BRITISH BOYS' LIBRARY.</u>

A New Series of 1s. 6d. books for boys.
Illustrated. Crown 8vo. Cloth extra.

***The Old Red School House.** A Story of the Backwoods. By Frances H. Wood.

***Ben**: A Story of Life's Byways. By Lydia Phillips, Author of "Frank Burleigh."

***The Secret of the Yew.** By Frank Yerlock.

***Major Brown**; or, Whether White or Black, a Man! By Edith S. Davis.

The Bell Buoy; or, The Story of a Mysterious Key. By F. M. Holmes.

Jack. A Story of a Scapegrace. By E. M. Bryant.

Hubert Ellerdale: A Tale of the Days of Wicliffe. By W. Oak Rhind.

<u>THE BRITISH GIRLS' LIBRARY.</u>

A New Series of 1s. 6d. books for girls.
Illustrated. Crown 8vo. Cloth extra.

***Regia**; or, Her Little Kingdom. By E. M. Waterworth and Jennie Chappell.

***Uno's Marriage.** By Mrs. Haycraft.

***Tephi**: An Armenian Romance. By Cecilia M. Blake.

***Christabel.** By J. Goldsmith Cooper, Author of "Nella."

Sweet Kitty Claire. By Jennie Chappell.

The Maid of the Storm: A Story of a Cornish Village. By Nellie Cornwall.

Mistress of the Situation (The). By Jennie Chappell.

Queen of the Isles. By Jessie M. E. Saxby.

<u>NEW PICTURE BOOKS.</u>

Happy and Gay: Pictures and Stories for Every Day. By D. J. D., Author of "Stories of Animal Sagacity," etc. With 8 coloured and 97 other Illustrations. Size 9 by 7 inches. Handsome coloured covers, paper boards with cloth backs.

Pleasures and Joys for Girls and Boys. By D. J. D., Author of "Anecdotes of Animals and Birds." With 8 coloured and 111 other Illustrations. Size 9 by 7 inches. Handsome coloured cover, paper boards and cloth back.

Anecdotes of Animals and Birds. By Uncle John. With 57 full-page and other Illustrations by Harrison Weir, etc. Fcap. 4to. 128 pages. Handsomely bound in paper boards, with Animal design in 10 colours, varnished. (A charming book for the Young.)

Stories of Animal Sagacity. By D. J. D. A companion volume to "Anecdotes of Animals." Numerous full-page Illustrations. Handsomely bound in paper boards, with Animal subject printed in 10 colours, varnished.

1s. 6d. each.

ILLUSTRATED REWARD BOOKS.

Crown 8vo. 160 pages. Cloth extra. Fully Illustrated.

*The Legend of the Silver Cup. Allegories for Children. By Rev. G. Critchley, B.A. With 12 Illustrations. (Small quarto.)

Aileen; or, "The Love of Christ Constraineth Us." By Laura A. Barter, Author of "Harold ; or, Two Died for Me."

Duff Darlington; or, An Unsuspected Genius. By Evelyn Everett-Green. With six Illustrations by Harold Copping.

Everybody's Friend; or, Hilda Danvers' Influence. By Evelyn Everett-Green, Author of "Barbara's Brother," etc.

Fine Gold; or, Ravenswood Courtenay. By Emma Marshall, Author of "Eaglehurst Towers," etc.

Jack's Heroism. A Story of Schoolboy Life. By Edith C. Kenyon.

Marigold. By L. T. Meade, Author of "Lady of the Forest," etc.

Nella; or, Not My Own. By Jessie Goldsmith Cooper.

Our Duty to Animals. By Mrs. C. Bray, Author of "Physiology for Schools," etc. Intended to teach the young kindness to animals. Cloth, 1s. 6d. ; School Edition, 1s. 3d.

Raymond and Bertha: A Story of True Nobility. By L. Phillips, Author of "Frank Burleigh ; or, Chosen to be a Soldier."

Rose Capel's Sacrifice; or, A Mother's Love. By Mrs. Haycraft, Author of "Like a Little Candle," "Chine Cabin," etc.

Satisfied. By Catherine M. Trowbridge.

Sisters-in-Love. By Jessie M. E. Saxby, Author of "Dora Coyne," "Sallie's Boy," etc. Illustrated by W. Rainey, R.I.

Ted's Trust; or, Aunt Elmerley's Umbrella. By Jennie Chappell, Author of "Who was the Culprit?" etc.

Thomas Howard Gill: His Life and Work. By Eliza F. Pollard, Author of "Florence Nightingale," etc.

Tamsin Rosewarne and Her Burdens: A Tale of Cornish Life. By Nellie Cornwall.

---

*Our Exemplar; or, What would Jesus do? By Charles M. Sheldon, Author of "The Crucifixion of Philip Strong," etc. 320 pages. Stiff paper Covers, 1s. 6d. Cloth boards, 2s. *This remarkable book has already had a sale in America of 185,000 copies.*

The Crucifixion of Phillip Strong. By Charles M. Sheldon, Author of "Our Exemplar," etc. Stiff paper covers, 1s. 6d. ; cloth boards, 2s.

*Insects: Foes and Friends. By W. Egmont Kirby, M.D., F.L.S., with Preface by W. F. Kirby, F.L.S., F.E.S., of the Natural History Museum, South Kensington. Demy 16mo. 32 pages of coloured Illustrations and 144 pages of descriptive letterpress. Cloth boards, 1s. 6d.

ONE SHILLING REWARD BOOKS.

Fully Illustrated. 96 pages. Crown 8vo. Cloth extra.

*Dumpy Dolly. By E. M. Waterworth, Author of "Master Lionel," "Lady Betty's Twins," etc.

*A Venturesome Voyage. By F. Scarlett Potter, Author of "The Farm by the Wood," etc.

*The Pilgrim's Progress. By John Bunyan. 416 pages. 47 Illustrations.

Always Happy; or, The Story of Helen Keller. By Jennie Chappell, Author of "Ted's Trust."

Birdie's Benefits; or, A Little Child Shall Lead Them. By Ethel Ruth Boddy.

Band of Hope Companion (The). A Hand-book for Band of Hope Members: Biographical, Historical, Scientific, and Anecdotal. By Alf. G. Glasspool.

Carol's Gift; or, "What Time I am Afraid I will Trust in Thee." By Jennie Chappell, Author of "Without a Thought," etc.

Brave Bertie. By Edith Kenyon, Author of "Jack's Heroism," "Hilda; or, Life's Discipline," etc.

Children of Cherryholme (The). By M. S. Haycraft, Author of "Like a Little Candle," "Chine Cabin," etc.

Cared For; or, The Orphan Wanderers. By Mrs. C. E. Bowen, Author of "Dick and his Donkey," etc.

Farm by the Wood (The). By F. Scarlett Potter, Author of "Phil's Frolic," etc.

Frank Burleigh; or, Chosen to be a Soldier. By L. Phillips.

Frank Spencer's Rule of Life. By J. W. Kirton, Author of "Buy Your Own Cherries."

Grannie's Treasures, and How They Helped Her. By L. E. Tiddeman.

His Majesty's Beggars. By Mary E. Ropes, Author of "Bel's Baby," etc.

Harold; or, Two Died for Me. By Laura A. Barter.

Jack the Conqueror; or, Difficulties Overcome. By the Author of "Dick and his Donkey."

Jenny's Geranium; or, The Prize Flower of a London Court.

1s. each.

POPULAR SHILLING SERIES.

Crown 8vo, well printed on good paper, and bound in attractive and tasteful coloured paper covers. Fully Illustrated.

Cousin Mary. By Mrs. Oliphant.

Louie's Married Life. By Sarah Doudney.

The Strait Gate. }
Grandmother's Child, and For Lucy's Sake. } By Annie S. Swan.

Living it Down. By Laura M. Lane.

Eaglehurst Towers. By Mrs. Emma Marshall.

Without a Thought. }
Her Saddest Blessing. } By Jennie Chappell

Fine Gold; or, Ravenswood Courtenay. By Emma Marshall.

The above can also be had in fancy cloth, price 1s. 6d.

CHEAP REPRINTS OF POPULAR STORIES FOR THE YOUNG.

Crown 8vo. 160 pages. Illustrated. Cloth boards, 1s. each.

*Claire; or, A Hundred Years Ago. By T. M. Browne, Author of "Jim's Discovery," etc.

*The Minister's Money. By Eliza F. Pollard, Author of "True unto Death," etc.

*Nobly Planned. By M. B. Manwell, Author of "Mother's Boy," etc.

*Her Two Sons. A Story for Young Men and Maidens. By Mrs. Charles Garnett.

Rag and Tag: A Plea for the Waifs and Strays of Old England. By Mrs. E. J. Whittaker.

Through Life's Shadows. By Eliza F. Pollard.

The Little Princess of Tower Hill. By L. T. Meade.

Clovie and Madge. By Mrs. G. S. Reaney.

The Best Things. }
Rays from the Sun. } By Dr. Newton

Ellerslie House: A Book for Boys. By Emma Leslie.

Manchester House: A Tale of Two Apprentices. By J. Capes Story.

Like a Little Candle; or, Bertrand's Influence. By Mrs. Haycraft.

Violet Maitland; or, By Thorny Ways. By Laura M. Lane.

Martin Redfern's Oath. By Ethel F. Heddle.

Dairyman's Daughter (The). By Legh Richmond.

1s. each.

PICTURE BOOKS FOR THE YOUNG.

Fcap. 4to. With Coloured Covers, and Full of Illustrations.

*Ring o' Roses: Pictures and Stories for Little Folks. By Uncle Jack, Author of " Frolic and Fun," etc. Four Full-page coloured and numerous other Illustrations.

*Holiday Joys: Stories and Pictures for Girls and Boys. By C. D. M., Author of " Merry Playmates," etc. Four full-page coloured and numerous other Illustrations.

Frolic and Fun: Pictures and Stories for Every One. By Uncle Jack, Author of " Follow the Drum," etc. Four full-page coloured and numerous other Illustrations.

Merry Playmates: Pictures and Stories for ' Little Folks. By C. D. M., Author of " Brightness and Beauty," etc. Four full-page coloured and numerous other Illustrations.

Follow the Drum: Pictures and Stories for Cheerful and Glum. By Uncle Jack, Author of " Bright Beams and Happy Scenes," etc. Four full-page coloured and numerous other illustrations.

Off and Away: Pictures and Stories for Grave and Gay. By C. D. M., Author of " Brightness and Beauty," etc. Four full-page coloured, and numerous other Illustrations.

Bible Pictures and Stories. Old Testament. By D. J. D., Author of " Pets Abroad," etc. With Forty-four full-page Illustrations. Coloured paper boards, 1s. ; cloth gilt, 1s. 6d.

Bible Pictures and Stories. New Testament. By James Weston and D. J. D. With Forty-five beautiful full-page Illustrations by W. J. Webb, Sir John Gilbert, and others. New Edition. Fcap. 4to. Illustrated boards, 1s.. ; cloth, extra, 1s. 6d.

Bright Beams and Happy Scenes: A Picture Book for Little Folk. By J. D. Four full-page coloured and numerous other Illustrations. Coloured paper cover, 1s. ; cloth, 1s. 6d.

Holiday Hours in Animal Land. (New Series.) By Uncle Harry. Four full-page coloured and numerous other Illustrations. Coloured paper cover, 1s. ; cloth, 1s. 6d.

Merry Moments. A Picture Book for Lads and Lasses. By C. D. M. Four full-page coloured and many other Illustrations. Coloured paper cover, 1s.; cloth, 1s. 6d.

BOOKS BY REV. DR. NEWTON.

New and Cheap Edition. 160 pages. Crown 8vo. Prettily bound in cloth boards, 1s. each.

Bible Jewels. | Bible Wonders.
Rills from the Fountain of Life.
The Giants, and How to Fight Them.
Specially suitable for Sunday School Libraries and Rewards.

*Molly and I. By the Author of " Jack." " At Sunset," etc. Long 8vo. Illustrated Title Page. 1s.

Cicely's Little Minute. By Harvey Gobel. Long 8vo. Illustrated Title Page. Cloth extra. 1s.

1s. each.

*Uncrowned Queens. By Charlotte Skinner, Author of "Sisters of the Master." Small 8vo. 112 pages. Cloth.

Sisters of the Master. By Charlotte Skinner, Author of "The Master's Gifts to Women."

The Master's Gifts to Women. By Mrs. Charlotte Skinner. Small 8vo. 112 pages. Cloth.

The Master's Messages to Women. By Mrs. Charlotte Skinner. (Uniform with the above.)

Some Secrets of Christian Living. Selections from the "Seven Rules" Series of Booklets. Small 8vo, cloth boards.

Daybreak in the Soul. By the Rev. E. W. Moore, M.A., Author of "The Overcoming Life." Imperial 32mo. 144 pages. Cloth.

Steps to the Blessed Life. Selections from the "Seven Rules" Series of Booklets. By Rev. F. B. Meyer, B.A. Small Crown 8vo, cloth boards.

Thoroughness: Talks to Young Men. By Thain Davidson, D.D. Small Crown 8vo. Cloth extra.

Women of the Bible. (Old Testament). By Etty Woosnam. Third Edition. Royal 16mo. Cloth.

9d. each.

NINEPENNY SERIES OF ILLUSTRATED BOOKS.

96 pages. Small Crown 8vo. Illustrated. Handsome Cloth Covers.

*Rob and I; or, By Courage and Faith. By C. A. Mercer.

*Phil's Frolic. By F. Scarlett Potter.

*How a Farthing Made a Fortune; or, Honesty is the Best Policy. By Mrs. C. E. Bowen.

A Flight with the Swallows. By Emma Marshall.

Babes in the Basket (The); or, Daph and Her Charge.

Bel's Baby. By Mary E. Ropes.

Benjamin Holt's Boys, and What They Did for Him. By the Author of "A Candle Lighted by the Lord."

Ben's Boyhood. By the Author of "Jack the Conqueror."

Ben Owen: A Lancashire Story. By Jennie Perrett.

Cousin Bessie: A Story of Youthful Earnestness. By Clara Lucas Balfour.

Dawson's Madge; or, The Poacher's Daughter. By T. M. Browne, Author of "The Musgrove Ranch," etc.

Five Cousins (The). By Emma Leslie.

9d. each.

Foolish Chrissy; or, Discontent and its Consequences. By Meta, Author of "Noel's Lesson," etc.

For Lucy's Sake. By Annie S. Swan.

Giddie Garland; or, The Three Mirrors. By Jennie Chappell.

Grandmother's Child. By Annie S. Swan.

How Paul's Penny Became a Pound. By Mrs. Bowen, Author of "Dick and his Donkey."

How Peter's Pound Became a Penny. By the same Author.

Jean Jacques: A Story of the Franco-Prussian War. By Isabel Lawford.

John Oriel's Start in Life. By Mary Howitt.

Left with a Trust. By Nellie Hellis.

Letty; or, The Father of the Fatherless. By H. Clement, Author of "Elsie's Fairy Bells."

Master Lionel, that Tiresome Child. By E. M. Waterworth.

Man of the Family (The). By Jennie Chappell.

Mattie's Home; or, The Little Match-girl and her Friends.

Paul, A Little Mediator. By Maude M. Butler.

Sailor's Lass (A). By Emma Leslie.

6d. each.

NEW SERIES OF SIXPENNY PICTURE-BOOKS.

Crown quarto. Fully Illustrated. Handsomely bound in paper boards, with design printed in Eight colours.

*Dollies' Schooltime: Pictures and Stories in Prose and Rhyme.

*Birdie's Message: The Little Folks' Picture Book.

Sweet Stories Retold. A Bible Picture-Book for Young Folks.

After School.

Doggies' Doings and Pussies' Wooings.

Little Snowdrop's Bible Picture-Book.

This New Series of Picture Books surpasses, in excellence of illustration and careful printing, all others at the price.

6d. each.

THE "RED DAVE" SERIES.

New and Enlarged Edition, with Coloured Frontispieces. Hand-somely bound in cloth boards.

*Joe and Sally; or, A Good Deed and its Fruits.
*The Island Home. By F. M. Holmes.
*Chrissy's Treasure. By Jennie Perrett.
*Puppy-Dog Tales. By Various Authors.
Mother's Boy. By M.B.Manwell.
A Great Mistake. By Jennie Chappell.
From Hand to Hand. By C. J. Hamilton.
That Boy Bob. By Jesse Page.
Buy Your Own Cherries. By J. W. Kirton.
Owen's Fortune. By Mrs. F. West.
Only Milly; or, A Child's Kingdom.
Shad's Christmas Gift.
Greycliffe Abbey.

Red Dave; or, What Wilt Thou have Me to do?
Harry's Monkey: How it Helped the Missionaries.
Snowdrops; or, Life from the Dead.
Dick and his Donkey; or, How to Pay the Rent.
Herbert's First Year at Bramford.
Lost in the Snow; or, The Kentish Fisherman.
The Pearly Gates.
Jessie Dyson.
Maude's Visit to Sandybeach.
Friendless Bob, and other Stories.
Come Home, Mother.
Sybil and her Live Snowball.
Only a Bunch of Cherries.
Daybreak.
Bright Ben: The Story of a Mother's Boy.

THE MARIGOLD SERIES.

An entirely new and unequalled series of standard stories, printed on good laid paper. Imperial 8vo. 128 pages. Illustrated covers with vignetted design printed in EIGHT COLOURS. Price 6D. each, NETT.

Pride and Prejudice. By Jane Austen.
From Jest to Earnest. By E. P. Roe.

The Wide, Wide World. By Susan Warner.

4d. each.

THE TINY LIBRARY.

Books printed in large type. Cloth.

Little Chrissie, and other Stories.
Harry Carlton's Holiday.
A Little Loss and a Big Find.
What a Little Cripple Did.
Bobby.
Matty and Tom.

The Broken Window.
John Madge's Cure for Selfishness.
The Pedlar's Loan.
Letty Young's Trials.
Brave Boys.
Little Jem, the Rag Merchant.

4d. each.

NEW FOURPENNY SERIES

of Cloth-bound Books for the Young. With Coloured Frontispieces. 64 pages. Well Illustrated. Handsome Cloth Covers.

Poppy; or, School Days at Saint Bride's.
Carrie and the Cobbler.
Dandy Jim.
A Troublesome Trio.
Perry's Pilgrimage.

Nita; or, Among the Brigands.
The Crab's Umbrella.
Sunnyside Cottage.
Those Barrington Boys.
Two Lilies.
Robert's Trust.

CHEAP "PANSY" SERIES.

Imperial 8vo. 64 pages. Many Illustrations. Cover printed in Five Colours.

*The Strait Gate. By Annie S. Swan.
*Mark Desborough's Vow. By Annie S. Swan.
*Her Saddest Blessing.
Miss Priscilla Hunter, and other Stories.
Wild Bryonie.
Avice. A Story of Imperial Rome.
A Young Girl's Wooing.
Spun From Fact.
A Sevenfold Trouble.
From Different Standpoints.
Those Boys.
Christie's Christmas.
Wise to Win; or, The Master Hand.
Four Girls at Chautauqua. }
The Chautauqua Girls at Home. }
Ruth Erskine's Crosses. }

Ester Ried. }
Julia Ried. }
Ester Ried yet Speaking. }
An Endless Chain.
Echoing and Re-echoing.
Cunning Workmen.
Tip Lewis and His Lamp.
The King's Daughter. }
Wise and Otherwise. }
Household Puzzles. }
The Randolphs. }
Mrs. Solomon Smith Looking On.
Links in Rebecca's Life.
Interrupted.
The Pocket Measure.
Little Fishers and their Nets.
A New Graft on the Family Tree.
The Man of the House.

3d. each.

THE PRETTY "GIFT-BOOK" SERIES.

With Coloured Frontispiece, and Illustrations on every page. Paper boards, Covers printed in Five Colours and Varnished, 3d.; cloth boards, 4d. each.

My Pretty Picture Book.
Birdie's Picture Book.
Baby's Delight.
Mamma's Pretty Stories.

Tiny Tot's Treasures.
Papa's Present.
Pretty Bible Stories.
Baby's Bible Picture Book.

BOOKS BY CHAS. M. SHELDON,

Author of "In His Steps," etc., etc.

OUR EXEMPLAR;

OR,

WHAT WOULD JESUS DO?

(IN HIS STEPS.)

Cloth boards, gilt edges, 2s. 6d. ; cloth boards, 2s. ; paper boards, 1s. 6d.
Special Edition in art linen, 1s.

"No one can read it without realising how far we are behind the great Example. Few will read it without being fired with the resolve to walk in His steps."— Rev. J. CLIFFORD, M.A., D.D.

THE CRUCIFIXION OF PHILLIP STRONG.

Cloth boards, gilt edges, 2s. 6d. ; cloth boards, 2s. ; paper boards, 1s. 6d. ;
Special Edition in art linen, 1s.

A powerful story of self-abnegation and its fruits.

HIS BROTHER'S KEEPER.

Cloth boards, gilt edges, 2s. 6d. ; cloth boards, 2s. ; paper boards, 1s. 6d.
Special Edition in art linen, 1s.

A stirring narrative on the great theme of Christian responsibility.

ROBERT HARDY'S SEVEN DAYS:

A Dream and Its Consequences.

160 pages, cloth boards, 1s.

A sure energiser of vigorous Christian effort.

RICHARD BRUCE.

Cloth boards, gilt edges, 2s. 6d. ; cloth boards, 2s. ; paper boards, 1s. 6d.
Special Edition in art linen, 1s.

A story of earnest Christian effort for the good of others by word, deed, and pen.

THE TWENTIETH DOOR.

Cloth boards, gilt edges, 2s. 6d. ; cloth boards, 2s. ; paper boards, 1s. 6d.
Special Edition in art linen, 1s.

MALCOLM KIRK.

224 pages, cloth boards, 1s.

A thrilling story of consecrated effort in a frontier settlement.

A Cheap Edition of the above books, in paper covers for distribution, can also be had, price 6d. each.

8 & 9, *PATERNOSTER ROW*, E.C.

www.ingramcontent.com/pod-product-compliance
Lightning Source LLC
Chambersburg PA
CBHW021802190326
41518CB00007B/412